SNOWDROP

Reaktion's Botanical series is the first of its kind, integrating horticultural and botanical writing with a broader account of the cultural and social impact of trees, plants and flowers.

Published

Apple Marcia Reiss
Ash Edward Parker
Bamboo Susanne Lucas
Berries Victoria Dickenson
Birch Anna Lewington
Cactus Dan Torre
Cannabis Chris Duvall
Carnation Twigs Way
Carnivorous Plants Dan Torre
Cherry Constance L. Kirker and Mary Newman
Chrysanthemum Twigs Way
Geranium Kasia Boddy
Grasses Stephen A. Harris
House Plants Mike Maunder
Lily Marcia Reiss
Mulberry Peter Coles
Oak Peter Young
Palm Fred Gray
Pine Laura Mason
Poppy Andrew Lack
Primrose Elizabeth Lawson
Rhododendron Richard Milne
Rose Catherine Horwood
Snowdrop Gail Harland
Sunflowers Stephen A. Harris
Tulip Celia Fisher
Weeds Nina Edwards
Willow Alison Syme
Yew Fred Hageneder

SNOWDROP

Gail Harland

REAKTION BOOKS

For Richard Fennell, even though he still doesn't know his nivalis from his elwesii, and for Ashley and Jonathan, of course.

Published by
REAKTION BOOKS LTD
Unit 32, Waterside
44–48 Wharf Road
London N1 7UX, UK

www.reaktionbooks.co.uk

First published 2016
Paperback edition first published 2024

Printed and bound in India by Replika Press Pvt. Ltd

A catalogue record for this book is available from the British Library

ISBN 978 1 78914 850 3

Contents

Snowdrops (*Galanthus nivalis* 'Flore Pleno') with cyclamen (*Cyclamen coum*).

Introduction

> What Spring have we? Turn back! –
> Though this be winter's end,
> Still may far-memoried snowdrops bloom
> For us, my friend.
>
> WALTER DE LA MARE, 'Blow, Northern Wind' (1950)

Miguel de Cervantes, whose best-known work *Don Quixote* (1605–15) is considered the first modern European novel, found that writing the introduction was the most difficult part of his book. The same could probably be said when writing a book on snowdrops, because of course snowdrops surely need no introduction. Those elegant white flowers that bloom at the tail end of winter have featured in paintings and poetry through many centuries and are familiar even to non-gardeners.

Snowdrops are one of the first flowers to open each year, the white nodding bells making an effective contrast to the yellow buttercup-like blooms of winter aconites or to the pink and magenta shades of cyclamen and the blues of reticulate irises. As individual flowers, snowdrops have an intricate beauty that demands that you kneel before them to better examine the distinctiveness of the green markings on the inner petals. But it is perhaps the wild or naturalized colonies, where snowdrops bloom together in their thousands, that attract the most attention. Those fortunate days when the low winter sunlight illuminates a carpet of white and the flowers release

Naturalized snowdrops at Painswick Rococo Garden, Gloucestershire, 2014.

their honeyed scent are to be treasured as one of the great joys of the winter months.

Although snowdrops are thought of as essentially late winter-flowering plants by most people, there are a number of different species or cultivars that will flower in the late summer or autumn and several whose season stretches from late autumn through to the turn of the year. By selecting a variety of species and cultivars, a gardener can have snowdrops in flower for possibly six to eight months of the year. At least in the northern hemisphere, though, it is the February-flowering snowdrops that are most valued by snowdrop lovers, and gardens with good collections can receive thousands of visitors in a day when they open at this time of year.

Snowdrops in British woodlands have the appearance of being wild, but they are generally accepted by experts to be not native but introduced plants that have become naturalized. Aaron Davis, author of *The Genus 'Galanthus'* (1999), points out that so-called wild populations of snowdrops in the British Isles are often composed of sterile plants, unlike those in their native habitats which set fertile seed.[1] Most colonies are associated with human habitation and they are frequently found around former monastic sites. The common snowdrop *Galanthus nivalis* is a vigorous plant that spreads rapidly by offsetting bulbs, or less frequently by seed, so that large colonies can form in just a few years. Rodents and rabbits tend to find the bulbs distasteful but may redistribute them when digging. Snowdrops are common in river valleys and in areas prone to flooding the bulbs may be further distributed by water.

There is some debate around the first use of the common name 'snowdrop'. The British botanist Arthur Church, writing in 1908, was of the opinion that the name was not old and was derived from the German *Schneetropfen* (snow drop), referring to the tear-shaped pearl earrings popular in the sixteenth and seventeenth centuries.[2] Church, a talented botanical artist who worked at the University of Oxford Botanic Garden, produced wonderfully detailed studies of flowering plants. Pendant earrings of the type he referred

Arthur Harry Church, *Galanthus nivalis*, 1904, watercolour.

to were certainly commonly featured in portraits of the time, perhaps most famously in Johannes Vermeer's *Girl with a Pearl Earring* (*c.* 1665), although other examples include the Milanese *Portrait of a Lady* of 1515 by Bernardino Luini, Titian's *Venus of Urbino* of 1538 and the *Lucrezia de' Medici* painted in 1560 by Agnolo di Cosimo, known as Il Bronzino. It was not just women who wore pearl earrings, as demonstrated by the portrait of Sir Walter Raleigh (1588) by an unknown English artist held in the National Portrait Gallery in London. 'Dingle-dangle', an alternative English common name for snowdrop, reflects the way the flowers dangle from the stem in the same manner as an earring from an ear lobe.

In the first edition of John Gerard's *Herball; or, Generall Historie of Plantes* (1597) the snowdrop is referred to as the 'timely flowring bulbus violet'. By the 1633 edition, revised by Thomas Johnson, however, the word 'snowdrop' is used. The next recorded use of the term 'snowdrop' is in 1664 in the book *Experiments and Considerations Touching Colours* by Robert Boyle. Boyle was an Irish chemist and physicist, best known today for Boyle's law, which concerns the pressure and volume of a gas at constant temperature. He recorded

Unknown artist, *Sir Walter Raleigh*, 1588, oil on panel.

Johannes Vermeer, *Girl with a Pearl Earring*, c. 1665, oil on canvas.

an experiment on the changes of the colour of jasmine flowers and snowdrops affected by the salts that remain when vegetable matter is burnt, describing 'those purely white flowers that appear about the end of winter, and are commonly called snowdrops'. Also in 1664 the English writer and gardener John Evelyn described in his *Kalendarium hortense* the flowers that were at their prime in his garden during December, which included yuccas and 'snow flowers or drops'. Evelyn is best known as the author of *Sylva; or, A Discourse of Forest Trees* (1664) and as a garden designer. His garden at Sayes Court, in Deptford in southeast London, was visited by Charles II in 1663. Evelyn could be described as an early environmentalist in that he was concerned about air pollution in London and suggested the planting of trees to purify the air.

The word 'snowdrop' was certainly familiar by the eighteenth century, when one of the first appearances of the snowdrop in poetry occurred, in Thomas Tickell's verse 'Kensington Gardens' (1722): 'A flower that first in this sweet garden smiled / To virgins sacred, and the Snow-drop styled.' There are a number of different English vernacular names which may well pre-date the use of the word 'snowdrop'. The English novelist Maria Louise Ramé, who was born in Bury St Edmunds, Suffolk, to an English mother and a French father, was described by the Irish poet William Allingham as having 'a voice like a carving knife' but was a successful writer of more than forty novels.[3] She wrote under the pseudonym Ouida and her book *Strathmore*, published in 1865, opens with a reference to 'White Ladies', 'a pretty old English name for snowdrops'. She seems more familiar with the word 'snowdrop', though, and in *Folle-Farine* (1871) describes a 'Sister of Charity, with a fair Madonna's face, bent above a little pot of home-bred snowdrops, with her tears dropping on the white heads of the flowers.'

Other English names

Other English names, such as fair maids of February and Candlemas bells, are associated with the Christian festival of Candlemas, which occurs forty days after Christmas, around 2 February. The name *fiore della purificazione* (purification flower) in Italy and the French *violette de la chandeleur* (violets of Candlemas) have the same derivation. In French, snowdrops are more commonly referred to as *perce-neiges*, or snow piercers; the French term was used in madrigals of the seventeenth century.[4] In French, the noun was originally considered to have a feminine gender but is now increasingly used in the masculine. The Pennsylvanian poet Catharine Harbeson Waterman Esling wrote in 1839 about Louise Cortambert's *La Langage des fleurs*, saying,

'It has been aptly termed by her countrymen *Perce neige* . . . and is with equal propriety called snow-drop in America.'[5] The Spanish name is *campanilla de invierno*, or little bell of winter. In the Netherlands snowdrops are called *sneeuwklokjes* and in Germany *Schneeglöckchen*, both of which mean 'snow bells'. An alternative German name is *Amselblume*, which translates as 'blackbird flower' and is a reference to their flowering time, which usually coincides with the period in which the blackbird starts to sing. This is usually in late January to early February, although older birds tend to start singing later.[6] The eighteenth-century British poet Thomas Tickell referred to snowdrops as 'vegetable snow', and certainly colonies of snowdrops in a woodland can look very like drifts of snow.

According to *The Handbook of Folklore*, published in 1913 by the Folklore Society in London, it was a common country belief that snowdrops should not be brought into one's house, 'as they will make the cows' milk watery and affect the colour of the butter'. Whether or not the cows could really be affected by flowers in the house, it was considered unlucky to decorate a room with cut snowdrops. The name death's flower relates to an old belief that a solitary snowdrop indicates impending death, with suggestions that it was inauspicious to bring snowdrops indoors. There were reports that 'snowdrops will bring a parting if brought indoors as a cut flower, but will bring happiness if outside under the windows in their beds.'[7]

The scientific name for the common snowdrop is *Galanthus nivalis* and in March 2012 the word 'galanthophile' was added to the *Oxford English Dictionary*. The word, however, which is used to refer to a person who collects, or at least loves, snowdrops, is not new. It is thought to have been coined by Edward Augustus Bowles (1865–1954), a botanical artist and author of several horticultural books including a monograph on crocuses and colchicums. He produced a useful review of the genus *Galanthus*, which he published in the *Journal of the Royal Horticultural Society* in 1918, and just before his death he contributed a chapter to Frederick Stern's book *Snowdrops and Snowflakes* (1956).

Snowdrops in Somersetshire, from *The Garden* (June 1878).

Bowles gardened at Myddelton House in Enfield in north London, and exchanged plants with many notable gardeners of the time. Bowles corresponded with Oliver Evelyn Penfold Wyatt (1898–1973) and started one letter to him with the words: 'Dear Galanthophil'.[8] Wyatt served as a lieutenant in the Royal Field Artillery in the First World War and in 1919 he was commended for gallantry and coolness while under machine gunfire. In 1933 he purchased Maidwell Hall to the north of Northampton and established a school there; known as Beak to his pupils, Wyatt developed the grounds into a garden notable for its carpets of spring bulbs. He was a successful breeder of lilies and also raised many different snowdrops, including the tall, strongly marked *Galanthus elwesii* 'Maidwell L' and the cultivar 'Kite', which on established bulbs often produces two flowers per stem.[9]

'Snow piercer' – a snowdrop emerging through snow.

Wyatt may have been the first documented galanthophile, but he was just one member of an ever-increasing group of people fascinated by these lovely plants. Early snowdrop devotees included James Allen of Shepton Mallet in Somerset, who in the 1880s deliberately raised snowdrop seedlings and is thought to have been the first to do so. Over a hundred of these were named, but few have survived to the current time, although 'Merlin', with its all-green inner segments, and 'Magnet', which has an unusually long pedicel, are still justifiably popular.

Samuel Arnott (1852–1930) of Dumfries, another early galanthophile, took up gardening after having to give up training to be a minister and his work in the family bakery due to ill-health. He regularly wrote about snowdrops and produced a pamphlet reviewing the genus.[10] A painting showing Arnott in his role as Provost of Maxwelltown is held at the Dumfries Museum and he is commemorated by the vigorous honey-scented snowdrop cultivar 'S. Arnott'. 'S. Arnott' is thought to be a hybrid between *Galanthus nivalis* and *G. plicatus*. Arnott sent bulbs to Henry John Elwes for him to trial in his Gloucestershire garden, Colesbourne Park, where now tens of thousands of bulbs grow in what is probably the largest display of this cultivar in the world.

James Atkins (1806–1884) was a nurseryman at Kingsthorpe Road in Northampton. He grew a diverse range of plants including pelargoniums and tropical orchids in his heated glasshouse.[11] He retired to a cottage on the Painswick estate in Gloucestershire, which is now famous for its snowdrops. Atkins became interested in alpine plants and in 1852 spent six weeks travelling in the Alps; in around 1870 he obtained a snowdrop bulb from somewhere around Naples in southern Italy which he considered to be *Galanthus imperati*. It proved to be a particularly large, early blooming snowdrop with elongated flowers and was released for sale around five years later. The identification was questioned and in 1891 it was renamed 'Atkinsii' to commemorate the man who introduced it into the nursery trade.

Modern galanthophiles are a varied bunch of people and include botanists, nurserymen, keen gardeners and an eclectic mixture of other people who just happen to be interested in snowdrops or want something to do on a wintry day. For the galanthophile, there are open gardens to visit, lectures and study days and plant sales to attend, and trips to visit plants in the wild to organize. And outside the main flowering season the addict can scan online auction sites for snowdrop memorabilia or visit art galleries to spot snowdrops in paintings. There are of course a proportion of people who are totally obsessed with the plants – who want to possess every species and cultivar and drive their family and friends mad with discussions on the minute differences in shades of green and white – but on the whole it seems to be a fairly harmless addiction.

Galanthophiles today are sometimes contemptuously dismissed as the 'twitchers' of the horticultural world. The garden writer

Galanthus 'Atkinsii', named to commemorate nurseryman and snowdrop enthusiast James Atkins.

Three kinds of snowdrop from *The Garden*, vol. XI (1877).

Christopher Lloyd (1921–2006) could usually be relied upon to express a controversial statement. He pointed out that 'galanthophiles can easily become galanthobores.' Lloyd himself, who gardened at his home Great Dixter in Sussex, grew around twenty different kinds of snowdrop, but wrote that 'a genuine nutter might have upwards of 300 and still be far from sated.'[12] The British writer and garden designer Mirabel Osler, writing in 1993, was more sympathetic: 'It's

Snowdrops for sale at the 2009 Galanthus Gala.

not hard to understand these obsessive galanthophiles; the flowers they are seeking have a furtive fragility which appeals to our quieter moods.'[13]

Not everyone is a fan of snowdrops. Elizabeth Barrett Browning wrote to Robert Browning in February 1845: 'To me unhappily, the snowdrop is much the same as the snow – it feels as cold underfoot.'[14]

Reginald Farrer, a plant explorer and writer who died in 1920 aged just forty while on a botanical expedition to Burma, wrote in his book *In a Yorkshire Garden* (1909):

> The Snowdrop gives me chilblains, only to look at it – and the very sight of a Snowdrop will always make me hurry to the fireside. Was there ever such an icy, inhuman, bloodless flower, crystallised winter in three gleaming petals and a green-flecked cup?

This is a view that would not be understood by the environmental campaigner Heather Tanner, who wrote in *Woodland Plants* (1981), 'Is there a creature so calloused that he has no special feeling for the snowdrop? Herein there is great hope for man, whatever his general depravity.' Even Farrer had to admit that the poculiform snowdrop, which has cup-shaped flowers made up of equal-sized pure white petals, has a flawless beauty.[15]

Snowdrops are seemingly flowers of contradictions. They appear to be plants of great fragility and in freezing conditions the flower stems will lose their rigidity and flop flat to the ground. As soon as temperatures start to rise, however, the stems quickly stand erect again and the flowers open their petals to attract any passing insects. The pristine white flowers are emblems of purity and innocence and yet they inspire avarice in many keen collectors and have even attracted the attention of criminals. Bulbs of particularly desirable cultivars regularly fetch prices in excess of £200 ($300) on online auction sites. In 2012 £725 (around $1,100) was paid for a bulb of the yellow-flowered *Galanthus woronowii* 'Elizabeth Harrison'. This record price for a single bulb was beaten in February 2015 when a bulb of *G. plicatus* 'Golden Fleece' sold at online auction for £1,390. The high prices reached by many cultivars have led on numerous occasions to plants being stolen from notable collections.

Overleaf: Snowdrops at a Royal Horticultural Society show in London, 2014.

GALANTHUS
AUGUSTUS
SMALL
TALK
Award of
Garden Merit

Wild populations of snowdrops show a wide range of natural variability and many naturally occurring variants have been introduced into cultivation. Snowdrop collectors will invariably have a range of different species and cultivars flowering at the same time, which increases the chances of cross-pollination occurring among their collection and therefore the likelihood of new and interesting forms arising. In recent years there have been an increasing number of people deliberately cross-pollinating plants to raise new hybrids, and there has been an unprecedented rise in the number of new cultivars named, such that to try to estimate how many different snowdrops there are in cultivation today would be something of a fool's game. There are three official National Collections of snowdrops in Britain, under the auspices of Plant Heritage, the largest of which in Bedfordshire has around 900 different taxa, although this is probably less than half the total number extant at the current time.[16]

Some collectors are now realizing that it is impossible to build up a complete snowdrop collection and so will specialize in one area, such as those with yellow markings, double flowers or green-marked outer petals. Britain was previously at the forefront of the collecting and naming of snowdrops, but many other countries are now catching up. The first snowdrop conference was held by the Royal Horticultural Society in London in 1891. In modern times, tickets for the Galanthus Gala, organized by Joe Sharman of Monksilver Nursery in Cambridgeshire, were avidly sought and the event was attended by snowdrop fanciers from as far away as Ohio and Japan. As a tribute to Her Majesty Queen Elizabeth II on her Diamond Jubilee, the people of Shaftesbury in Dorset planted thousands of snowdrops in public areas around the town to create a series of 'snowdrop walks'. Shaftesbury is now the home of an annual Snowdrop Festival that incorporates talks, a bulb sale and an art exhibition.

Many other countries are now organizing similar events, which can include lectures, visits to snowdrop collections, and snowdrop swaps or sales which attract long queues of people hoping to be able to buy particularly desirable forms. In Belgium snowdrop days are held at the Arboretum Kalmthout, and in the Netherlands a Sneeuwklokjesfeest (Snowdrop Festival) has been held at De Boschhoeve in Wolfheze since 2001. In Germany, snowdrop days were organized at the Oirlicher Blumengarten in Nettetal by the author and plantsman Günter Waldorf until his death in 2012.

There are no snowdrops native to North America, but gardeners there are also becoming infected with the snowdrop obsession. Indeed, Barbara Tiffany, writing for the Mid-Atlantic Group of the Hardy Plant Society, warned, 'This is a very dangerous genus. For no apparent reason, people lose their minds over these tiny things. Proceed with caution.'[17]

Snowdrops (*Galanthus nivalis* 'Flore Pleno') with spring snowflakes (*Leucojum vernum*) on the Cambo Estate in Fife.

Snowdrops (*Galanthus nivalis* 'Flore Pleno') naturalized on the Cambo Estate in Fife.

one

Among Trees and Rocks

> I love being chaste now. I love it as snowdrops love the snow.
>
> D. H. LAWRENCE, *Lady Chatterley's Lover* (1928)

Snowdrops are usually thought of as primarily woodland plants, and certainly the common snowdrop (*Galanthus nivalis*) can be seen carpeting woods from the Pyrenees to the Balkans. Although it is true that they tend to prefer cool growing conditions in soils that are moist during the flowering season, the genus *Galanthus* contains twenty species which are found in a wide diversity of habitats. In the alpine meadows of the high Caucasus Mountains, *G. platyphyllus* grows at altitudes of up to 2,600 m (8,500 ft), where it experiences long, cold winters but plenty of light and moisture while in growth. On the Greek island of Kastellorizo (Megísti), *G. peshmenii* grows in rocky crevices, surprisingly close to the seashore. *G. fosteri* has an even more southerly distribution, growing on shady ledges and outcrops in Turkey, Syria and Lebanon.

Snowdrops are members of the Amaryllidaceae, the large family of plants that includes familiar garden bulbs such as daffodils (*Narcissus*) and popular houseplants such as *Clivia* and *Hippeastrum*. The family takes its name from the South African genus *Amaryllis*, which is named after a beautiful shepherdess featured in Virgil's pastoral work *Eclogues* of the first century BC. A genus is defined as a group of closely related species, whereas a species is perhaps more difficult to define. As expressed in the *Royal Horticultural Society Shorter Dictionary of Gardening*, a species consists of 'closely related, morphologically similar individuals, often found within a distinct geographical range'. The name for the snowdrop genus, *Galanthus*, was derived by Linnaeus from the Greek words *gala*, meaning milk, and *anthos*, flower. The names of other genera of plants have similar derivations, including the pea relative *Galega*, which was once thought to increase the milk flow of nursing mothers, and *Galium*, whose name is a reference to the former use of the plant *Galium verum*, or lady's bedstraw, to curdle milk in the manufacture of cheese.[1] *Galanthus*, by contrast, is simply a reference to the milk-white colour of the flowers.

Within the Amaryllidaceae family, *Galanthus* is most closely related to the genera *Acis* and *Leucojum*. These two genera were formerly united within the genus *Leucojum* and differ from *Galanthus* in that each flower has six equal-sized petals. They are commonly known as snowflakes and, like a true snowflake, the flowers exhibit a sixfold radial symmetry. The genus *Leucojum* was split into two in July 2004.[2] The two species with hollow stems, and broader leaves and petals marked with green or yellow, remain in *Leucojum*; these are the spring snowflake, *Leucojum vernum*, which is often mistaken for a snowdrop, and the taller, later-flowering summer snowflake, *L. aestivum*. Those nine species which are characterized by narrow leaves, solid flower stems and flowers without markings have been reclassified as belonging to *Acis*. These include the easy-to-grow autumn snowflake *Acis autumnalis* and the sugar-pink, fairy-like *A. rosea*, which is found growing wild in Corsica and Sardinia where it flowers in late summer and early autumn. *Acis* species tend to grow in drier areas than those favoured by snowflakes and snowdrops.

This reclassification of *Leucojum* species was not a totally new idea. Indeed, the genus *Acis* was first defined by the British botanist Richard Anthony Salisbury in 1807, in his text to William Hooker's *Paradisus Londinensis* (1805–8) where he remarked that the leaves 'are so narrow as to give me some suspicion that it may constitute a distinct genus from the broad-leaved *Leucojums*'.[3] Salisbury did not agree with the 'sexual system' for classifying plants used by Carl Linnaeus, favouring instead the natural system of classification, particularly as used by the Frenchman Antoine Laurent de Jussieu. Salisbury's work was ignored by many of his fellow botanists at the time, who accused him of plagiarism, and *Acis* was retained as a separate genus until 1878. At that time, specimens of the plants were reviewed by John Gilbert Baker (1834–1920), an assistant curator of the herbarium at the Royal Botanic Gardens at Kew. Baker was born in Guisborough in North Yorkshire and had attended Bootham School, a Quaker school established in 1823 that has a renowned natural history society, thought to be the oldest of its kind in Britain. Baker worked in his father's grocery business in Thirsk and was active in the local botanical society. In 1864 the family business and home were lost in a fire, after which he took up the position at Kew. Baker concluded that *Acis* were not sufficiently distinct from *Leucojum* species to warrant their own genus.[4] This remained the accepted view until a detailed study by M. Dolores Lledó and colleagues in 2004 concluded that *Acis* was a valid genus after all. The name *Acis* comes from Ovid's *Metamorphoses*, where it is the name of the shepherd boy who loved the sea nymph Galatea and was murdered by the jealous cyclops Polyphemus.

Today the naming of plants is covered by a set of rules known as the International Code of Botanical Nomenclature.[5] This sets out how to correctly name plants and also prescribes what to do should two or more names be found to have been used accidentally for the same species. In such cases, the code stipulates that the earlier name must be used, in order to give due recognition to the first person to name that species. Changes in the names of plants can occur for other reasons, which usually relate to more extensive study of the plants in question. In particular, modern DNA analysis can reveal previously unsuspected relationships between plants, and sometimes necessitate a plant being moved from one genus to another or even between different families. Today new plants are still being discovered and there are developments in the understanding of the geographical range of particular species and in their natural variabilities.

Galanthus have perhaps suffered less from name changes than many other genera. However, even a supposedly simple question such as 'how many snowdrop species are there?' can lead to arguments between botanists. There are differences of opinion; for example, botanists in Georgia recognize *G. schaoricus* as a distinct species, whereas others consider this to be the same plant as *G. alpinus*.

What is a Snowdrop?

Snowdrops may be confused with snowflakes and also possibly from a distance with certain other white-flowering bulbs, but the genus *Galanthus* is really quite a homogeneous group. All its species are small herbaceous perennials that grow from bulbs. They have strap-like leaves and usually one white flower per stem. Snowdrops are in leaf for an average duration of four to four and a half months each year.

The leaves of snowdrops are always paired. The arrangement and colour of the leaves is important for identification of different species. The arrangement of the leaves of a plant in bud is known as vernation, and in snowdrops it is possible to examine the vernation in mature plants by looking at how they are folded at the base of the plant. In applanate vernation, as seen in the common snowdrop, the two leaves are pressed flat to each other. Explicative leaves, also known as plicate or pleated, are also flat against each other, but the edges of the leaves are folded back or sometimes rolled. The Crimean snowdrop, *G. plicatus*, has this type of leaf. In supervolute plants such as *G. elwesii*, one leaf is clasped around the other within the bud and generally remains so at the point where the leaves emerge from the soil.[6]

As the flowering stem, or inflorescence, emerges from the ground the flower bud is enclosed in a spathe, which consists of two bract-like structures connected by a transparent membrane. This protects the young bud and breaks as the bud matures. The snowdrop's nodding flowers dangle from a slender pedicel and are shaped like little bells. Simple flowers such as the buttercup have green sepals which

Snowdrops with explicative leaves: *G. plicatus*.

Left: Supervolute leaves on *G. elwesii*.

Right: Snowdrops with applanate leaves: *Galanthus* 'South Hayes'.

protect the flower bud and the decorative petals within. In flowers such as tulips the sepals have developed to look like the petals and cannot be clearly distinguished from them; the term 'tepal' is usually applied to these structures. In *Galanthus*, however, the petal-like structures are variously known as petals, tepals or perianth segments, usually shortened to segments. Snowdrops have three larger outer segments surrounding three smaller green-marked segments that form a little inverted 'cup'. There is a small notch at the apex of each inner segment, the function of which is not clear.

Snowdrops are generally cross-pollinated, although they will sometimes self-pollinate. They are specifically adapted for pollination by bees. The green markings on the inner segments serve as a sign that there is nectar within while also achieving efficient photosynthesis and so providing the flower and developing seeds with additional nutrients.[7] The flowers are usually scented, often with a sweet, honey-like fragrance. The nodding nature of the flowers protects the pollen from rain. Opening of the flowers is temperature-dependent, with the outer petals lifting at temperatures over 10°C (50°F).[8] Snowdrop pollen is a bright-orange colour and is very attractive to bees. The individual pollen grains are oblong with a single long groove; the grains measure 19–34 μm by 17–27 μm.[9]

Snowdrop seed pods usually ripen in early summer, changing from green to yellow and finally brown. The seeds are round and creamy white when first shed, but they are comparatively immature at this point and the embryo continues to develop. The endosperm contains a plentiful supply of starch to fuel growth and development. Seeds have a limited tolerance of drying out and develop best when kept in moist conditions at 20°C (68°F).[10] In wild populations, the viability of snowdrop seeds has been assessed as 52–60 per cent.[11] For every 10 grams (0.35 oz) of seed, there are approximately 1,000 individual seeds. Each seed has a small, hooked tail known as an elaiosome, positioned on the opposite side to the embryo. This is an

Most snowdrop flowers have three longer outer segments and three inner segments marked in green, as in this flower of *Galanthus* 'Silverwells'.

Snowdrop capsules and seeds with oil-rich elaiosomes.

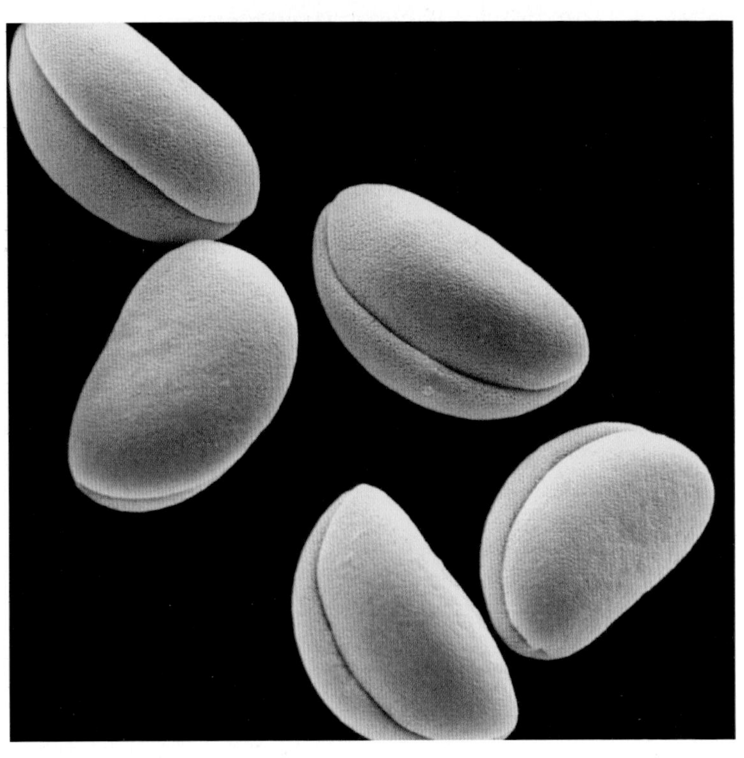

Enhanced scanning electron micrograph of snowdrop (*Galanthus nivalis*) pollen.

outgrowth of the seed with large oil-rich cells containing substances attractive to ants. Elaiosomes are common in Eurasian species of the Amaryllidaceae family and serve to encourage insects to distribute the seeds.

Snowdrop seeds start to germinate as the weather begins to cool in September. The primary root and sheath of the seed leaves emerge and grow rapidly for four to five days in order to establish the young plant within the leaf litter. Growth then slows and the sprouting seed can overwinter in this state. Seeds within the same population will continue to germinate in the spring. Snowdrops have just one seed leaf, a type of germination (monocotyledonous) that is also found in many other plant families, such as grasses, lilies and orchids. Distinguishing between plants with one seed leaf or two (dicotyledons) is a key step in classifying a plant; a distinction first made by the Cambridge academic and naturalist John Ray, in his book *Methodus plantarum nova* of 1682.

Calcium oxalate crystals have been found in the young roots, leaves and flowers of G. *plicatus*, G. *gracilis* and G. *elwesii*.[12] The crystals are known as raphides and are needle-shaped. Similar structures have been found in more than 200 families of plants but their role is still unclear, though they are most likely a defence mechanism against browsing animals.

Studying Snowdrops

One of the first botanical illustrations of a snowdrop occurs in *De historia stirpium commentarii insignes*, which was published in 1542.[13] The author, Leonhart Fuchs, is commemorated in the plant genus *Fuchsia*. His book, a hefty folio-sized volume, was written to review all those plants known to have recognized medicinal properties. It is illustrated with over five hundred plates of plants, drawn from nature by Albrecht Meyer. Over one hundred species of plants were illustrated for the first time in Fuchs's book; many of the engravings were based on specimens that came from Fuchs's own garden in Tübingen,

Het tvveede Deel. Het ſevenſte Boeck. 361

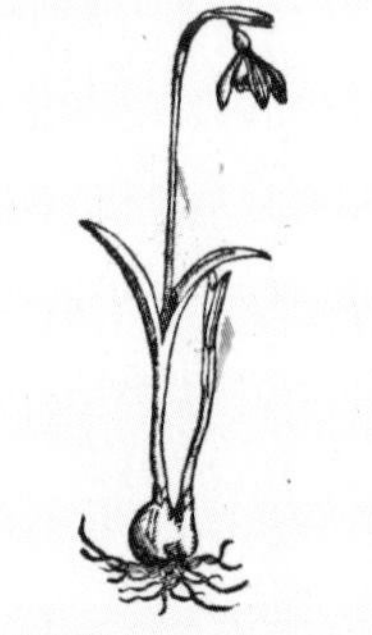

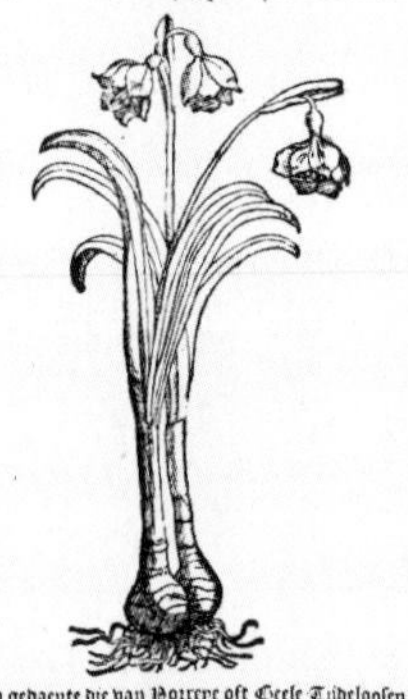

HET XXVI. CAPITEL.

Van Witte Tijdeloosen oft *Leucoion bulboſum.*

Gheſlachten.

The snowdrop in Rembert Dodoens's *Herbarius oft Cruydteboeck* (1644 edition).

near Stuttgart in Germany. The plants illustrated include a spring snowflake (*Leucojum vernum*) and a snowdrop, although the snowdrop specimen looks as though it was rather past its best.

In Italy in 1554 the physician and naturalist Pietro Andrea Mattioli produced his major work, *Medici senensis commentarii, in sex libros pedacii Dioscoridis*. This contained an illustration of a many-leaved snowdrop which was labelled as a *Narcissus*, as was usual at that time. Snowdrops, of course, have only two leaves, and so the artist may have been inadvertently combining characteristics of the *Leucojum* and *Galanthus* genera.

Rembert Dodoens was a Flemish physician and botanist (also known by his Latinized name, Rembertus Dodonaeus) who in 1554 published his herbal *Cruydteboeck*, influenced by Fuchs's herbal. Dodoens likewise concentrated on those plants that were used medicinally. The book was a great success and in its time it was the most translated book after the Bible. It was used by the London barber-surgeon and plant collector John Gerard as the source of much of his famous *Herball; or, Generall Historie of Plantes* of 1597, with additional descriptions of plants such as the potato that had been newly introduced from America. Gerard's herbal included *L. vernum*, *L. aestivum* and *G. nivalis*.

The genus name *Galanthus* was used by Linnaeus in *Systema naturae* (1735), and in *Species plantarum* (1753) he described the common snowdrop, *G. nivalis*. The first classification of *Galanthus* was carried out by Pierre Edmond Boissier in 1882, when six species were recognized.[14] Six years later John Gilbert Baker described the same species and recognized two subspecies of *G. nivalis*.[15] In 1956 Sir Frederick Stern produced a major study of snowdrops and snowflakes, which was published by the Royal Horticultural Society.[16] It became the standard reference for these plants but has been criticized for its lack of information regarding taxonomic studies on populations of wild snowdrops that were carried out in the former USSR.[17]

The Botanical Magazine Monograph on Galanthus, written for the Royal Botanic Gardens, Kew, by Aaron Davis in 1999 and beautifully illustrated by Christabel King, was the next major work on the genus and did much to fuel the increase in interest in it. This surge in snowdrop popularity further developed following the publication in 2001 of the book known to galanthophiles as 'the bible'. *Snowdrops: A Monograph of Cultivated Galanthus* by Matt Bishop, Aaron Davis and John Grimshaw is considered essential reading for snowdrop enthusiasts. A second edition to cover recently introduced cultivars is forthcoming.

The Species

Galanthus alpinus was described in 1911 by the Georgian botanist and mycologist Dmitrii Ivanovich Sosnowsky based on plants that he collected from Mt Lomis Mta in Georgia. Sosnowsky co-authored with A. Makaschvii a flora of Georgia in 1941 and is commemorated in the name of the hogweed *Heracleum sosnowskyi*. The name *G. alpinus* now encompasses the plants formerly called *G. caucasicus* and *G. bortkewitschianus*. The species grows in the Caucasus and Transcausasus, northeastern Turkey and possibly in northern Iran.[18] It is found at a wide range of altitudes, usually in or at the margins of deciduous forests. It is a dainty species with quite narrow leaves. There are two varieties of *G. alpinus*. The first, var. *alpinus*, has whitish bulb scales and fertile seed capsules, whereas var. *bortkewitschianus*, found in Chegem in southern Russia, has yellowish bulb scales and sterile seed pods. *G. alpinus* is not widely grown and seems slow to bulk up in gardens.

Plants labelled *G. angustifolius* in Britain usually turn out to be narrow-leafed forms of *G. nivalis*. The true species *G. angustifolius* is thought to be cultivated in only a few specialist collections.[19] It was first described by Jurij Ivanovich Koss in 1951 from collections made in the northern Caucasus, where it grows in mixed deciduous woodland or scrub in deep clay soils.[20] It experiences very cold winters in its wild environment. Plants have narrow glaucous (covered in a bluish-grey waxy coating), applanate leaves. The flowers are similar to those of *G. nivalis*, with a simple U- or V-shaped green mark; forms with green-flushed outer petals have been found in the wild.

First described by John Gilbert Baker in 1897, *G. cilicicus* was named after the area of Asia Minor between the Taurus Mountains and the Mediterranean that was known in classical times as Cilicia and corresponds to the modern region of Çukurova in Turkey. *G. cilicicus* is found in nature among limestone rocks, growing in pockets of humus within the limestone. It is notable for its flowering time, which

Galanthus alpinus: a wild plant near Novorossiysk, Russia.

Galanthus angustifolius in the northern Caucasus, 2014.

extends from autumn into winter. The horticulturalist Graham Stuart Thomas reported that his specimens often flowered in December.[21] The plant is closely related to G. *peshmenii*, which also flowers in autumn, but the leaves of G. *cilicicus* are several centimetres long at flowering whereas those of G. *peshmenii* are only just emerging at flowering time – although at maturity they are actually longer than those of G. *cilicicus*. The flowers have a single green mark on each inner perianth segment. G. *cilicicus* is known to grow at only a few localities in the wild and is also rare in cultivation. It is considered to be difficult to grow, being less robust than other species and less inclined to bulk up well. Some growers report that it does best grown in a pot under glass.[22]

G. *elwesii* was described by Graham Stuart Thomas as 'the most handsome snowdrop in foliage', and it certainly does have attractive, broad leaves, which have a bluish-grey bloom to them.[23] The leaves are supervolute, with one firmly wrapped around the other at the base. It is usually a large plant with good-sized flowers that are strongly marked, although very small forms are also known. The markings vary widely: some plants have two separate marks, one at the apex and one at the base of the inner segments; others have a single large mark that is roughly X-shaped but may cover most of the surface, or there can be a U-shaped apical mark. Plants with just a single mark are known as G. *elwesii* var. *monostictus*. The species was named in 1875 after Henry John Elwes by Sir Joseph Dalton Hooker, the director of the Royal Botanic Gardens at Kew.[24] G. *elwesii* grows wild across parts of the Balkans and Ukraine. It grows mainly in mountain regions and prefers north-facing slopes, which remain cool in summer. In cultivation it is an easy to grow plant that flowers freely, although recently imported wild-collected bulbs can be unreliable.

G. *fosteri* is a distinctive species named to honour Professor Michael Foster, who imported bulbs of the species from the Amasya region of northern Turkey. Foster (1836–1907) was born in Cambridgeshire. He was a physiologist with a strong interest in irises, and created numerous iris hybrids which were instrumental in the development of bearded iris cultivars. The Foster Memorial Plaque is awarded yearly by the British Iris Society for exemplary service in the work of iris culture. G. *fosteri* is found mainly in central and southern Turkey and in Jordan, Syria and Lebanon, where it grows in shady places at altitudes between 1,000 and 1,600 m. The supervolute leaves are a deep, shiny green. Its flowers always have both apical and basal markings; the apical marking is shaped like a horseshoe but is often almost cut in half by the notch at the tip. The species grows best if allowed a warm, dry rest in summer. Its mature bulbs often split into smaller ones, which then take a year off to build in size before flowering, although some clones will flower regularly each year.[25]

G. *gracilis* is closely related to G. *elwesii* but is usually a smaller plant with narrower leaves. The specific name *gracilis* comes from the Latin for 'graceful' or 'slender' and is an appropriate name for the plant. Wild plants usually grow to about 10 cm in height and the flowers are usually attractively marked. They grow in a range of habitats at altitudes from around 100 m up to around 2,000 m, often alongside ivy-leaved cyclamen (*Cyclamen hederifolium*). They are found in

Galanthus fosteri, illustration by M. E. Eaton in the journal *Addisonia* (1931).

Bulgaria, northeastern Greece, Romania, Ukraine and western Turkey. The species was first named by the Prague-born botanist Ladislav Josef Čelakovský in 1891, based on herbarium specimens from Bulgaria. Čelakovský studied the Czech flora in great detail and published many new species. He is commemorated in the names of several plants, including the orchid *Orchis celakovskyi*. The leaves of *G. gracilis* often have a pronounced twist to them, an effect that is particularly noticeable in the cultivar 'Corkscrew'. This form is thought to have been found by Ken Aslet, who was working on the rock garden at the Royal Horticultural Society's garden at Wisley; it was introduced to commerce by the British firm North Green

Snowdrops in 1997.[26] G. *gracilis* 'Highdown', which originated at the South Downs garden of Sir Frederick Stern in Worthing, is the most commonly grown cultivar of *gracilis* in gardens. In cultivation, bulbs of G. *gracilis* are not always long-lived but they renew themselves readily from seed.[27]

G. *ikariae* was named after the Greek island Icaria, so-called because Icarus of Greek mythology was meant to have fallen into the sea nearby. G. *ikariae* is found only on Icaria and a few other Aegean islands. It grows, often alongside ivy-leaved cyclamen, mainly in humid wooded river valleys in quite deep shade, where conditions can be extremely wet at flowering time in February but are dry in summer, at which time the bulbs are protected from extreme heat by

Galanthus koenenianus cultivated in a terracotta pot.

bracken. First described by Baker in April 1893, G. *ikariae* is often confused with other green-leafed snowdrops, particularly G. *woronowii*; in fact G. *ikariae* has darker green leaves and a much larger mark on the inner segments of the flowers, usually covering over half of the tepal. In cultivation it seems less hardy than many other snowdrop species and does better in mild microclimates, as might be expected from its Greek homeland. A naturalized population was first reported in the UK in 1984, in Hertfordshire.[28]

G. *koenenianus* was discovered in northeast Turkey in 1988 and named after Manfred Koenen, the German plant collector who

Galanthus krasnovii growing in western Transcaucasia with *Pachyphragma macrophyllum*.

Galanthus lagodechianus in cultivation with crocus.

found it.[29] It grows in mixed forest at altitudes of 1,000–1,700 m. *G. koenenianus* is a very unusual snowdrop, being highly distinctive even out of flower due to its leaves, which are distinctly ribbed on the undersides. It is a dainty plant, reaching only around 9 cm tall in flower, with fairy-like, well-marked flowers. The leaves are short and clasp the stem at flowering so that the ribbing can easily be seen. The plant is said by some to have an unpleasant fragrance, although this does not always seem to be the case and the scent can be faint but sweet; it flowers in the early mid-season.

The uncommon species *G. krasnovii* grows in Georgia and northeast Turkey around the eastern Black Sea coast. It is found among rocks near the Chakvistskali river and in clearings in beech woodland. It

was formally described in 1963 by the Russian plant collector and botanist Andrej Pavlovich Khokhrjakov, although it had been recorded in herbarium specimens from the early twentieth century.[30] The name honours the Russian botanist Andrej Nikovaevich Krasnov (1862–1914). Like its near relative *G. platyphyllus*, *G. krasnovii* lacks the typical apical notch on the inner segments of the flower. The leaves of wild plants can be 30 cm long and 6 cm wide at maturity, with upper surfaces that are often slightly puckered.

The Lagodekhi National Reserve, established in 1912, is the oldest protected area in Georgia. It has given its name to the snowdrop *G. lagodechianus*, which was described in 1947 by the Georgian botanist Liubov Manucharovna Kemularia-Nathadze, who worked on the flora of Georgia and produced a review of the Caucasian species of peony.[31] *G. lagodechianus* is found in mixed deciduous forest in the mountains of the central and eastern Caucasus, in Armenia, Azerbaijan and Iran. Plants have narrow, shiny green applanate leaves and a single apical green mark on the inner segments. It is one of the last species to flower, often into April in habitat, and in cultivation can flower two weeks or more later than the similar species *G. rizehensis*.

G. nivalis is usually referred to as the common snowdrop, but although it is undoubtedly widespread in its natural distribution and in cultivation, it rivals all other species for beauty. In his book *Winter Blossoms from the Outdoor Garden* (1926) Anthony William Darnell reported that

> we have gathered the most exquisite blossoms we have ever seen of the common Snowdrop on the borders of woods on the south side of Dartmoor; in point of size they nearly equal those of *G. elwesii* and were far more graceful.

G. nivalis grows wild from the Pyrenees to the Ukraine and has been naturalized throughout much of northern Europe. It prefers

Galanthus nivalis, the common snowdrop.

deciduous woodland but may also be found among scrub and rocks, particularly in moist, humid places. The leaves of G. *nivalis* are applanate and usually a glaucous grey-green; the lower leaf surface is slightly greyer than the upper. Flowers have a single green mark at the apex of the inner segments. Plants of G. *nivalis* vary considerably in size, leaf characteristics and flower form. Variants such as those with green markings on the outer tepals, yellow markings, atypical segments or split spathes are not uncommon in wild populations, as seen in the widely grown cultivar 'Scharlockii', which was discovered in 1868 by the physician Julius Scharlock in the German Nahe valley.[32]

G. *panjutinii* is endemic to the chalky ridges of the northern Colchis area of Krasnodar territory, Russia. It was first reported in 1913 by G. Sakharov after it was found in a clearing above the town of Gagra in Abkhazia, growing with plants including the peony *Paeonia abchasica*. The first sample collected was given the specific epithet G. *valentinae* by the climber and naturalist Platon Sergeevich Panjutin, but it was never validly published by him and the species was not formally described until 2012.[33] The leaves are bright green with an oily sheen to them. The outer segments of the flower are long and bend outwards quite widely at maturity; its inner segments have a small apical mark which may be reduced to two dots. G. *panjutinii* grows in alpine meadows and forest under the Nordmann fir (*Abies nordmanniana*). A large part of one of five known locations for the plant was destroyed due to the construction of facilities for the 2014 Sochi Winter Olympics.

Named, like the green-flowered *Puschkinia peshmenii*, to commemorate the Turkish botanist Dr Hasan Peşmen (1939–1980), who died in a traffic accident, G. *peshmenii* was formally described in 1994.[34] It has a limited distribution in the Antalya province of southern Turkey and on the Greek island of Kastellorizo. It flowers in the autumn, usually before the leaves start to emerge, but differs from the better-known G. *reginae-olgae* in having leaves that are narrower and longer with a flaccid appearance at maturity. Under a microscope the leaves are clearly distinct from *reginae-olgae* as they have

Galanthus panjutinii from the northern Colchis area of western Transcaucasia.

a well-defined palisade layer of cells below the epidermis. Its closest relative is considered to be the rare *G. cilicicus*. There are several clones of *G. peshmenii* in cultivation; one, named 'Kastellorizo' after the island, is said to be the showiest, with freely produced flowers that are slightly larger than usual *peshmenii*.[35] *G. peshmenii* grows in crevices in rocks on north-facing cliffs by the sea, where it escapes predation by goats. In cultivation it grows well as a pot plant when protected from the worst of winter wet and cold in a sand plunge. In mild microclimates it can grow outdoors, particularly if planted in a raised bed or rock garden situation.

Previously known as *G. latifolius*, a name that has also been used for other similar species, *G. platyphyllus* was formally described in 1948 by the American botanists Hamilton Paul Traub and Harold Norman Moldenke. The name *platyphyllus* is of Greek derivation and means 'broad-leaved', which is an apt description of the plants since at maturity the leaves can be as much as 6 cm wide.[36] The leaves are a shiny, bright green and wrap around each other. Plants of this species have yellowish, elongated bulbs similar to those of the closely related species G. *krasnovii*. It is one of the latest-flowering species. G. *platyphyllus* flowers generally lack the typical snowdrop notch to the inner segments and have blunt-tipped anthers; there is also, usually, a single U-shaped apical mark on the inner segments, although this may sometimes be divided into two small triangular marks. G. *platyphyllus* is found in Georgia and southern Russia, mainly at high altitudes of more than 2,000 m. It usually grows in grasslands above the tree line where the ground is wet from melting snow. The species is rare in cultivation as it is quite slow to increase; the conservation of wild plants is a concern, since bulb collectors have been stripping natural populations in Georgia.[37]

In his *Rariarum stirpium per Pannonias observatorum historiae* (History of Rare Plants Observed in Pannonia) of 1583, the Flemish botanist Carolus Clusius reported receiving a bulb of G. *plicatus* from Madame de Heysentein in Constantinople, along with a quantity of daffodils. In pre-Linnaean times this species was usually called *Leucojum bulbosum praecox byzantinum*, although this name was also sometimes used for large forms of the common snowdrop.[38] It was formally named in 1819 by the German botanist and explorer Friedrich August Marschall von Bieberstein from plants collected in the Crimea. Von Bieberstein combined a military career with wide-ranging botanical and archaeological interests. He compiled the first comprehensive flora of the Caucasus, first published in French in 1798. His collection of some 10,000 herbarium specimens is held by the Komarov Botanical Institute of the Russian Academy of Sciences. The epithet *plicatus* refers to the plant's pleated leaf margins, which are folded flat towards the underside of the leaf. The subspecies *byzantinus* is distinguished by having flowers that have two marks on each inner segment, one at the apex and the other at the base.

G. *plicatus* has a patchy distribution around the western part of the Black Sea in parts of the Crimean peninsula from Balaclava to

Koktebel, Romania and northwestern Turkey. In Russia three populations are known in the Novorossiysk region of the Caucasus.[39] It is found in or close to mixed deciduous and coniferous forests on both lime and acidic soils. It is easy to grow in cultivation and builds up large colonies in a pleasing manner. The flowers are not always as large as those of other species but they are generally reliably produced. In his book *Hardy Flowers* the Irish gardener and journalist William Robinson (1838–1935) warned, 'It must not be supposed however that it is so pretty as the common snowdrop though more than this has often been claimed for it.' Many named cultivars of this species have been selected, including 'Warham', which was introduced from the Crimea in around 1855 during the Crimean War, and 'Colossus', a particularly large and vigorous snowdrop found at Colesbourne Park in Gloucestershire. The cultivar 'Trym', in which the outer three petals appear to have been replaced by an extra set of inner segments having the same strong markings, originated in the garden of Jane Gibbs of Westbury-on-Trym in Bristol. It was named by Chris Brickell of the Royal Horticultural Society at Wisley.[40] A yellowish marked seedling of 'Trym' has been named 'Golden Fleece'.

The autumn-flowering G. *reginae-olgae* was first collected in Greece on Mt Taygetus in the Peloponnese, where it is still quite abundant. It grows in and around woodland, especially that of the oriental plane, pine and fir trees, growing mainly in nooks between rocks when humus has gathered in shady, damp places. It was described in 1876 by the Greek botanist and poet Theodoros Georgios Orphanides and named in honour of Queen Olga of Greece (1851–1926), who was the grandmother of Prince Philip, the Duke of Edinburgh. She was a member of the Romanov dynasty but married George I of Greece at the age of sixteen. She returned to Russia after the assassination of her husband in 1913 but following the Russian Revolution spent the last years of her life in Britain, France and Italy. Orphanides studied in Paris and published his first book of poetry in 1836. He became professor of botany in Greece in 1848 and obviously valued his G. *reginae-olgae* snowdrops very highly: the German

Wild plants of *Galanthus plicatus* growing through leaf litter.

horticulturist Max Leichtlin (1831–1910) tried to purchase stock from Orphanides but 'the exorbitant price demanded prevented his doing so.'[41] Orphanides was later confined to a mental institution and his snowdrops disappeared from view.

There has been some disagreement as to whether G. *reginae-olgae* is a distinct species or just a variant of G. *nivalis*, but the botanist Aaron Davis is emphatic that it warrants specific status due to its autumn-flowering period and because its leaves are absent or only partially developed at flowering time and have a glaucous stripe down the midrib.[42] The subspecies *vernalis* differs in that it flowers from winter to spring and has leaves that are present at the onset of flowering.

Named for the town of Rize in northeastern Turkey, G. *rizehensis* was first collected in 1933 and was exhibited at the Royal Horticultural

Society by George Percival Baker the following year. Baker, who is commemorated by *Tulipa bakeri*, was head of a textile printing firm as well as a keen mountaineer and president of the Iris Society. *G. rizehensis* was formally described by Sir Frederick Stern in 1956. Its geographical distribution includes western Georgia and southern Russia, where it grows in coastal woodland and humid forest, usually in the shade of deciduous trees. Most of the plants in cultivation are thought to originate from those Turkish forms grown by Stern in his garden at Highdown in West Sussex. *G. rizehensis* plants have dull green leaves which recurve as the plant comes into flower. There is a single apical mark on the inner flower segments. *G. rizehensis* can be confused with the closely related species *G. lagodechianus*, but this latter

Georgios Jakobides, *Queen Olga of Greece*, 1915, in whose honour *Galanthus reginae-olgae* was named.

usually has leaves of a bright, shining green and is at least two weeks slower to flower. In cultivation the flowers of G. *rizehensis* often appear in January, while in the wild they may flower from January to March.[43]

Named for the Transcaucasian region in which it is found, the snowdrop G. *transcaucasicus* was described by the Russian botanist Aleksandr Vasiljevich Fomin on the basis of material collected in Azerbaijan.[44] Fomin (1869–1935) was director of the St Vladimir University Botanical Garden in Kiev, Ukraine. When he died the university renamed the garden the A.V. Fomin Botanical Garden in his honour. G. *transcaucasicus* also occurs on the Black Sea coast around Batumi and in northern Iran, growing in the mountains among and alongside deciduous woodland. The flat, deep-green supervolute leaves are quite broad and have a distinct oily sheen to their upper surface. The flower is pure white, with a small U-shaped mark or

Galanthus rizehensis is a small species first found in 1933 above Trebizond in Turkey.

just touched with green on the inner segments; it is held on a pedicel which is as least as long as the spathe. It is one of the least-studied species and is rare in cultivation. Bulbs from Iran often start to flower in November or December whereas the Azerbaijan form usually flowers in February. Plants referred to as G. *caspius* from the Talysh region of Azerbaijan are synonymous with this species.

G. *trojanus* is considered to be a critically endangered species. It is found only in the Çanakkale province of northwestern Turkey, and very few populations are known.[45] It was first discovered during a botanical survey of western Turkey in 1994 by a team of botanists from Istanbul and Kew. The seeds of an unidentified snowdrop were collected and raised in the alpine house of the Royal Botanic Gardens at Kew, where they flowered for the first time in 1997. The name comes from the ancient region of Troad, which included the city of Troy. G. *trojanus* grows in undisturbed Turkey oak (*Quercus cerris*) woodland at altitudes of 350–590 m alongside plants such as *Anemone blanda* and the beautiful red peony *Paeonia peregrina*. It is similar to G. *nivalis* and G. *rizehensis* but usually taller, with bigger flowers, and can be distinguished by the broader leaves which are matt green with a slight cast of grey on the upper surface; the lower surfaces are green and shiny. The flowers have a small inner mark which is restricted to the apical quarter of the segment. A number of alkaloids, including two with pharmacological potential identified in 2014, were isolated from the species.[46]

G. *woronowii* was first grown in Britain in the latter half of the nineteenth century, with most of the plants having been sourced in Georgia. The species was formally described in 1935 by the Russian botanist Agnia Losina-Losinskaja, who wrote botanical reviews of strawberries and rhubarbs. G. *woronowii* was named to commemorate Yury Nikolaevich Voronov (1874–1931), the plant collector and botanist at the Leningrad Botanical Garden.[47] Losina-Losinskaja based her description of the snowdrop on plants collected near Sochi in southern Russia. The species is also found in the Pontic Mountains of Turkey and in the western Caucasus: it is extremely common in the Batumi

Galanthus woronowii, wild near Sochi, at Krasnaya Polyana.

area of southwest Georgia, where it grows in a wide variety of habitats and varies considerably in both leaf and flower.[48] It is sometimes found growing as an epiphyte, rooting in moss on living or fallen tree trunks. Russian botanists recognize three different ecological forms, but given the natural variability in the species there is some disagreement about the validity of their conclusions. There has been much confusion over the naming of this plant and similar species with broad green leaves and flowers with a single apical mark. Millions of bulbs of this species were dug up from the wild from the 1990s onwards to supply the horticultural trade. International and national laws try to regulate the trade in wild-collected bulbs and local Turkish farmers raise some one million bulbs each year for sale. Clones vary in their tolerance to conditions in cultivation but are generally easy to grow.

Naturally occurring hybrids of snowdrops occur from time to time in places where two species grow together. *Galanthus* × *allenii* was named by John Gilbert Baker in 1891 in honour of James Allen, the well-known British snowdrop enthusiast. The hybrid is thought to derive from a cross between G. *alpinus* and G. *woronowii*, but the exact parentage is uncertain.[49] G. × *allenii* has greenish-grey leaves and a single V-shaped apical mark on the inner segments. The flowers have a distinctive almond scent. G. × *hybridus* covers those plants formerly called G. × *grandiflorus*, which arise from crosses between G. *elwesii* and G. *plicatus*. Such hybrids are generally vigorous plants and include 'Merlin', which has all-green inner segments, and the popular plant 'Robin Hood'. The name G. × *valentinei* covers all hybrids between G. *nivalis* and G. *plicatus*.[50] This cross has occurred many times in cultivation, giving rise to robust and beautiful hybrid plants such as 'Magnet' and the variable 'Mrs Thompson'.

two

Purity and Piety

Make Thou my spirit pure and clear
As are the frosty skies,
Or this first snowdrop of the year
That in my bosom lies.
ALFRED, LORD TENNYSON, 'St Agnes' Eve' (1837)

People love stories and often seem to need to invent a mythology around a given subject if one does not already exist. This can make it difficult to distinguish long-held associations from more modern inventions. The snowdrop, however, is a familiar and much-loved flower and certainly has had symbolic value for many hundreds of years at least. A Christian legend has it that the snowdrop originated as a gift to Eve after she and Adam were expelled from the Garden of Eden for tasting the forbidden fruit. As Eve sat crying in the desolate wasteland beyond the garden and shivered as snowflakes began to fall all around her, an angel transformed the snowflakes into snowdrops to console her for the loss of her life in the garden and to give her hope for the future.

A version of this tale occurs in the poem 'Origin of the Snowdrop' by George Wilson (published in 1860), which suggests that the snowdrop, like the rainbow, is a heavenly promise of better things to come:

And thus the snowdrop, like the bow
That spans the cloudy sky,

Becomes a symbol whence we know
That brighter days are nigh.

Wilson (1818–1859) started life as a doctor and chemist and was appointed the first Regius Chair of Technology at the University of Edinburgh in Scotland in 1855. The author Margaret Oliphant, who was a second cousin of Wilson, reported that he was an 'excellent talker, full of banter and a kind of humour'.[1] Despite his scientific background, much of his writings were concerned with religious matters. 'The Sleep of the Hyacinth', subtitled 'An Egyptian Poem', was inspired by the report of a bulb found interred with an Egyptian princess that was said to have grown when it was planted. The poem wove scientific theories through with Wilson's reflections on life, death and resurrection.

With their demurely nodding, immaculate white flowers, it is perhaps not surprising that snowdrops became associated with purity and religion. In his poem 'A Snowdrop' (1929) the English poet and writer Walter de la Mare saw a manifestation of the Holy Trinity in the triple inner and outer petals: 'Beneath these ice-pure sepals lay, / A triplet of green-pencilled snow'. He tried to further his understanding of God through his contemplation of the flower. De la Mare's father's ancestors were French Huguenot silk merchants and his mother's family were Scottish evangelists, and he was once a choirboy at St Paul's Cathedral in London; but although he was steeped in the Christian tradition he had a dislike of dogma and many of his written works such as the novel *The Return* (1910) and the poem 'The Listeners' (1912) have a supernatural content.

Snowdrops are linked particularly with the Christian festival of Candlemas, which is celebrated on 2 February.[2] This of course coincides with the main snowdrop flowering season in much of the northern hemisphere. Few other plants are in flower at this early time of year. However, it is not just a coincidence of timing but also the physical appearance of the snowdrop flower that makes it an appropriate symbol for the festival. Candlemas is alternatively

WHEN ICE-FLAKES FELL IN SHOWERS
VPON THAT WORLD OF DEATH
SOME PITYING ONES DISTILLED
BY ANGEL-BREATH FELL LOV-
ING FLOWERS
AND SNOW-DROPS THVS WERE BORN
TO COMFORT EVE WHO SORROWFVL
THE LAND OF LIFE DID LEAVE

Frances MacDonald McNair, *The Legend of the Snowdrops*, 1900, watercolour and silver paint on linen.

known as the Feast of the Purification of the Virgin and commemorates the ritual purification of Mary forty days after the birth of Jesus, in accordance with Mosaic Law (Leviticus 12:2–8). It was believed that a woman was unclean after having given birth and that she needed to be ritually purified at her place of worship before rejoining her community. The day was also the occasion when the infant Jesus was presented at the Temple in Jerusalem and was met by Simeon and the prophetess Anna (Luke 2:22). Originally considered to be a Feast of the Lord, particularly in Eastern Orthodox countries, a ritual procession on the day was introduced by Pope Sergius I (687–701) and the festival became a celebration of Mary, with the term 'Candlemas' entering common use some time after the eleventh century.[3] It became the day on which all the church's candles for the year were blessed. Celebrations included processions in and around the church during which the participants hold lighted candles. There are many reports of regional variants, such as processions of girls dressed in white and the strewing of snowdrops on the altar, but documented evidence for such practices has not been identified in sources from the time.

Older, alternative names for snowdrops such as Candlemas bells, fair maids of February and white ladies have a connection with the Candlemas festival, conjuring up images of white-clad maidens. The name Mary's taper is a reference to the immature flower stem, which emerges slender and erect, like a long, thin candle. Echoes of these festivals lingered in some regions such as Herefordshire and Shropshire, where there was a tradition of bringing snowdrops into the house to cleanse it after the winter in an act of 'white purification'.[4] In *Chroma*, Derek Jarman's (1942–1994) meditative and very personal book written in the last year of his life, he reflected on the meaning of the colours white, red, blue and yellow: 'White is the colour of mourning except in the Christian West where it is black – but the object of mourning is white. Whoever heard of a corpse in a black shroud?' Jarman, a film director, artist and gardener, grew snowdrops in his Dungeness

garden and loved how they, with his emerging honeybees, heralded the coming of spring.[5]

Candlemas, like St Swithin's Day on 15 July, was a time of weather divination. Tradition has it that good weather on Candlemas indicates the return of severe wintry weather, as recorded in the rhyme:

> If Candlemas day is clear and bright,
> winter will have another bite.
> If Candlemas day brings cloud and rain,
> winter is gone and will not come again.

This is thought to originate from the Latin saying 'Si sol splendescat Maria purificante, major erit glacies post festum quam fuitante' (If the sun shines on the day of Mary's purification it will be icier after the feast than it was before). This was recorded in 1678 in *A Collection of English Proverbs* by the philosopher and naturalist John Ray.

In Germany it was said prior to the late 1800s that if the weather is sunny on Candlemas and a bear sees his shadow he will crawl back into his lair, assuming that there will be six more weeks of winter. This tradition was taken to the United States by German immigrants to Punxsutawney, Pennsylvania, where it developed into the Groundhog Day made famous worldwide by the comedy film of 1993 starring Bill Murray and Andie MacDowell. Sadly, groundhogs actually do not have proven abilities in weather forecasting. Similarly, snowdrops flowering on Candlemas may be a sign that spring is on its way but it does not necessarily mean that it will be arriving shortly.

An American winter event of a different kind takes place in Pottsville, a town in Arkansas which was named for Kirkbride Potts, who built a home at the base of Crow Mountain in the 1840s. The Greater Pottsville Winter Carnival was established in 1968 and

Walter de la Mare's 'triplet of green-pencilled snow' as shown here with G. 'Cowhouse Green'.

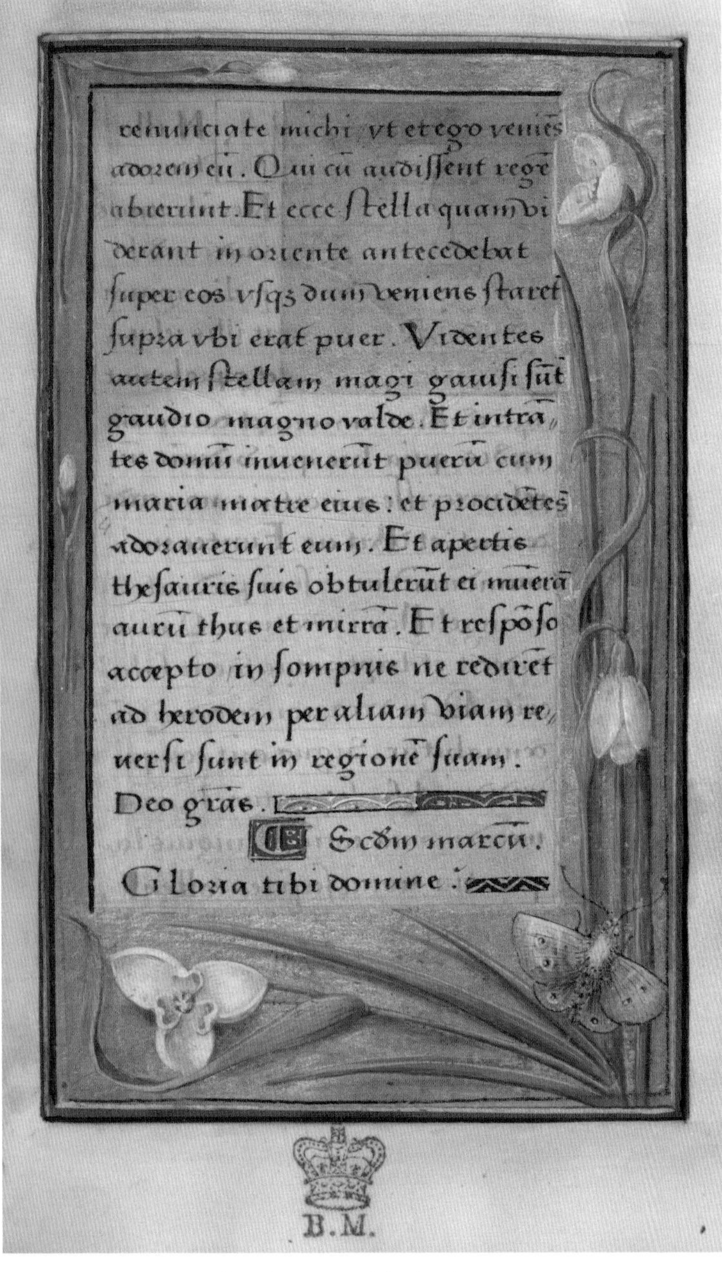
renunciate michi ut et ego veniēs
adorem eū. Qui cū audissent regē
abierunt. Et ecce stella quam vi
derant in oriente antecedebat
super eos usq; dum veniens staret
supra ubi erat puer. Videntes
autem stellam magi gavisi sūt
gaudio magno valde. Et intra
tes domū invenerūt puerū cum
maria matre eius: et procidētes
adoraverunt eum. Et apertis
thesauris suis obtulerūt ei muerā
aurū thus et mirrā. Et respōso
accepto in sompnis ne rediret
ad herodem per aliam viam re
versi sunt in regionē suam.
Deo grās.
Scdm marcū.
Gloria tibi domine.

Snowdrop from a French book of hours, *c.* 1500.

includes a pageant in which girls clad in white vie for the title of Snowdrop or Snowflake Princess.

Candlemas, like many Christian festivals, contains elements taken from other, older beliefs. It occurs at the midpoint of winter, halfway between the winter solstice and spring equinox, which was celebrated in pre-Christian times as a festival of light. Fires were lit to ward off evil spirits and to usher in the spring. Candlemas has also sometimes been linked with the ancient Roman feast of Lupercalia, in which men known as Luperci (the brothers of the wolf) purified the city before the New Year celebrations. This festival was observed around the fourteenth day of February, which later became associated with St Valentine. Today St Valentine's Day celebrations often involve presenting bunches of red roses to loved ones; however, while the idea of red roses is fairly entrenched in the minds of modern lovers, it is actually a relatively recent phenomenon, since importing out-of-season flowers has become possible only with cheaper airfreight. A snowdrop posy or even a potful of flowering snowdrop bulbs would perhaps be a more appropriate and romantic gesture than imported or greenhouse-grown roses. The snowdrops would certainly last longer and indeed if the bulbs were planted in the garden afterwards they could increase and flourish throughout the coming years.

Another saint sometimes associated with the snowdrop is St Agnes (*c.* 291–*c.* 304), who was martyred in Rome for her faith.[6] She is the patron saint of chastity, young girls and gardeners, and is generally shown with a lamb because her name resembles *agnus*, the Latin word for lamb. Her feast day is 21 January, when snowdrops may well be in flower. It is a time of year that was described evocatively by John Keats in the opening lines of his poem 'The Eve of St Agnes' (1819):

> St Agnes' Eve – Ah, bitter chill it was!
> The owl, for all his feathers, was a-cold;
> The hare limp'd trembling through the frozen grass,
> And silent was the flock in woolly fold . . .

In most depictions in art, St Agnes is shown accompanied not by snowdrops but rather by white lilies and a palm frond, which is used in Christian iconography to show a martyr's victory of spirit over flesh. Traditionally, it is said that on St Agnes Eve maidens will dream of their future lovers. They could also enact certain rituals to bring their lover to them, such as passing a piece of wedding cake through a wedding ring nine times, as illustrated in the painting *The Bridesmaid* (1851) by the Pre-Raphaelite artist Sir John Everett Millais.

Historically, Imbolc was a Gaelic festival that marked the beginning of spring.[7] It was celebrated in Ireland, Scotland and the Isle of Man and was considered to be a day for the literal spring cleaning of the home as well as a spiritual cleansing. The name is thought to derive from the Old Irish *i mbolc*, meaning 'in the belly',

Late Victorian period valentine card.

referring to the season when ewes are with lamb. The festival was associated with the Gaelic fertility goddess Brigid, who was considered to be the patroness of poetry, medicine and cattle. She was later Christianized as St Brigid, whose feast day is 1 February.

The popular snowdrop cultivar *Galanthus* 'Imbolc' was distributed from the garden of Primrose Warburg (1920–1996) at Yarnells Hill, Oxford. Warburg held annual snowdrop lunches for fellow galanthophiles at Imbolc.[8] She had a reputation for brusqueness but could be very generous. Guests at her Imbolc lunches were asked which snowdrops they particularly admired in her garden and often received a little package of bulbs later in the year.

Many apparently native colonies of snowdrops result from bulbs planted as memorials in churchyards alongside graves or at religious institutions. Snowdrops are frequently found around former monastic sites across Britain. The ruins of Walsingham Priory in Norfolk are surrounded by sheets of naturalized snowdrops that now attract many botanical pilgrims in late winter and early spring each year. The few remaining walls of the former Greyfriars Priory at Dunwich on the Suffolk coast enclose another colony. Most of the former city of Dunwich has been lost to the sea due to coastal erosion but the thirteenth-century Franciscan priory and the leper hospital of St James were built to the southwest of the city and escaped serious flooding in the late thirteenth and early fourteenth centuries. Ankerwyke Priory in Surrey, near the meadows of Runnymede, has many naturalized snowdrops and is also famous for the Ankerwyke Yew, a tree said to be at least 1,400 years old that is located in the grounds of a twelfth-century Benedictine nunnery.

Other snowdrop sites with religious connections include those at Forde Abbey in Somerset, Hodsock Priory in Nottinghamshire and the Cistercian Roche Abbey near Maltby in Yorkshire. Vigorous colonies such as those found at these sites were often used as a resource for local people who wanted to grow the flowers in their own gardens, and so bulbs have been spread around nearby villages.

Snowdrops were used to decorate churches and were often brought into the home to purify the house, though in some areas it has long been considered unlucky or a sign of death to bring snowdrops indoors on or before Candlemas. Richard Mabey suggests that this belief may have anti-Catholic roots, as with the similar superstition that bringing May blossom into the house is unlucky.[9] On overcast days when snowdrops hold their petals tightly together the blooms can suggest to the imaginatively minded person an array of miniature corpses in their shrouds, which may explain the link with death. The ability of bulbs to die back and then to produce new flowers again out of the seemingly barren earth also gives them a link to ideas about resurrection, and so they are frequently referenced in religious poetry. For example, Robert Buchanan wrote in his work *David Gray and Other Essays, Chiefly on Poetry* of 1868: 'Out of the snow, the snowdrop – Out of death comes life'.

Christina Rossetti, the poet and sister of the Pre-Raphaelite artist Dante Gabriel Rossetti, loved snowdrops and frequently took them to the grave of her sister Maria, who died of cancer in 1876. Christina associated the flower with the Virgin Mary, Maria's namesake. Rossetti's book *Called to be Saints* (1881) has a dedicatory page to Maria with a snowdrop decoration, and the book includes a commentary on the snowdrop, praising the flower for its 'coolness and purity, with refined humbleness and patience of hope'. Her poem 'Feast of the Presentation' is known in particular for its last verse:

> Then snowdrops and my heart
> I'll bring, to find those blacker than Thou art:
> Yet, loving Lord, accept us in good part;
> And give me grace to wait
> A bruised reed bowed low before Thy gate.

Rossetti's poignant 'Another Spring', written in 1862, also includes a reference to snowdrops:

If I might see another Spring
I'd not plant summer flowers and wait:
I'd have my crocuses at once,
My leafless pink mezereons,
My chill-veined snow drops, choicer yet
My white or azure violet . . .

The poem expresses the thought that we fail to appreciate the today in imagining that the future will be brighter, and it continues with a promise that if spared to live another spring the narrator would live life to the full and savour every experience, enjoying the snowdrops of today rather than dreaming of the summer flowers to come.

Victorian guides on etiquette encouraged the sending of flowers to the bereaved as a token of sympathy and in the hope of offering consolation. One such book published in 1891 quoted from the poem 'Bring Flowers' by the English Romantic poet Felicia Hemans (1793–1835): 'They are love's last gift – bring ye flowers, pale flowers!'

In Seamus Heaney's poem 'Mid-term Break', in which the poet described returning home from school to the funeral of his brother, 'Snowdrops and candles soothed the bedside.' Heaney, who won the Nobel Prize in Literature in 1995, wrote this extremely poignant and moving poem, which was published in his collection *Death of a Naturalist* in 1966, as an elegy to his younger brother Christopher, who died after being hit by a car at the age of four. The flowers and candles were used as a gesture of solace for a grieving family.

When Queen Victoria died on 22 January 1901 she was laid out at Osborne House on the Isle of Wight surrounded by lilies and by snowdrops gathered from the grounds, echoing those in the posy that she carried on marrying Prince Albert on 10 February 1840. The Queen had stipulated that there should be a white pall to cover her coffin and she was veiled with her Honiton lace wedding veil. The watercolour portrait *Queen Victoria on her Death Bed* by Sir Hubert von Herkomer, from a sketch he made of her on 25 January, is framed in ebony and on display at Osborne House outside the Queen's bedroom. It is difficult to distinguish the snowdrops in the painting, although other white flowers including arum lilies, lily of the valley and tuberose are recognizable. A wreath of snowdrops was sent to the funeral from the Royal Botanic Gardens at Kew.

The first commercial Christmas cards are thought to be those designed for Sir Henry Cole in 1843.[10] He was interested in industrial design and was the first director of the Victoria & Albert Museum in London, then called the South Kensington Museum. He commissioned the artist John Calcott Horsley to design an image of a family feasting, using the words 'A Merry Christmas and a Happy New Year to you', in order to send Christmas greeting to his friends and business acquaintances without having to write out a seasonal letter to each individual. The idea quickly caught on and designs such as ones of children pinning up garlands of holly and other seasonal plants were popular. One of the most well-known greetings card and postcard producers was Raphael Tuck & Sons, a business established in 1866 by Prussian-born Raphael Tuck and his wife, Ernestine, in Bishopsgate in the City of London. Their first Christmas card was produced in 1871. In 1880 the Tucks' son

The snowdrop symbolizes hope in the language of flowers, as on this postcard from the United States, *c.* 1909.

Raphael Tuck & Sons, snowdrops and canary on a Christmas card, Victorian era.

Adolph launched a competition with 5,000 guineas in prizes for the best Christmas card designs. This resulted in some 5,000 entries and created much publicity for the company. Snowdrops and the Christmas rose (*Helleborus niger*) were common images on Victorian Christmas and New Year cards, sending hopeful wishes for the coming year. They likewise often featured on Easter cards, although in the garden many snowdrops will have finished flowering by Eastertime.

Written valentine greetings have been exchanged for hundreds of years, with the first being traditionally attributed to Charles, Duke of Orléans, who was confined to the Tower of London after the Battle of Agincourt in 1415. He spent his time there writing romantic

Victorian snowdrop and flower bell Christmas card.

verses to his wife back in France. A Valentine's Day message written by Margery Brews to her fiancé John Paston in 1477 forms part of the Paston Papers kept at the British Library. Written greetings remained the norm until Victorian times, although homemade cards decorated with lace and silk flowers became popular. The oldest-known printed valentine, which is on display at York Castle Museum, was printed on 12 January 1797 by the publisher John Fairburn of 146 Minories, London.[11] It includes a verse printed around festoons of flowers, doves and cupids surrounding a seated woman:

> Since on this ever Happy day,
> All Nature's full of Love and Play

Yet harmless still if my design,
'Tis but to be your Valentine.

Old valentine cards with recognizable images of flowers most frequently depicted roses, blue forget-me-nots and pansies, though snowdrops and other spring flowers also appear. In the U.S. Esther Howland of Massachusetts is credited with starting the Valentine's Day card industry.[12] After her graduation from Mount Holyoke College in 1847 she received an English valentine from a business associate of her father and was so charmed by it that she imported paper lace and decorations from England in order to make some herself. She developed a small outlet for handmade cards into a thriving business. Many of the cards she made had delicate paper lace details and hand-painted silk or satin centres, or used intricate folding or lift-up flaps. Today the number of cards sent for St Valentine's Day is second only to those posted at Christmas.

In Denmark on Valentine's Day, or Fjortende Februar (Fourteenth of February), schoolchildren would send friendship tokens known as

May
Christmas
find you happy and leave you blest.

Victorian snowdrop Christmas card.

NEW YEAR'S GREETING.
A Flower, sweet friend I bring,
A gleam of light from clay,
That waiting not for Spring,
Shines forth on New Year's day!
As little, Oh, may we
Regard what season rolls,
So this sweet snow-drop be
No whiter than our souls!
FRANCIS DAVIS,

Victorian New Year card.

Danish *gaekkebrev*, with snowdrops; the number of dots indicates the number of letters in the writer's name.

gaekkebrev to each other. The *gaekkebrev* is a humorous love note written on paper that is then cut into a lacy pattern. A pressed snowdrop flower would be included with the note. Custom dictates that the notes are not signed but include a number of dots, one for each letter in the name of the author.[13] If the recipient correctly guesses the identity of the person who sent the note they can claim an Easter egg from them on Easter Sunday. If they fail to guess then they must pay a forfeit. Forerunners to the *gaekkebrev* in the 1600s were letters sealed in complicated puzzle knots, or a series of knots on a piece of string enclosed in an envelope with a letter, the letters beautifully decorated with flowers and verses. If you were unable to undo the knot you had to give a party. Even in these days of texting and social media, a survey commissioned for the Danish Post Office revealed that 19 per cent of Danes expect to receive a valentine on 14 February.

The popular use of in memoriam cards followed on from the introduction of commercial Christmas cards, although they were first recorded in the Netherlands in the seventeenth century. The Museum of Religious Art in Uden in the province of Brabant in the Netherlands has tens of thousands of examples of prayer cards,

Victorian Easter card.

devotional pictures and in memoriam cards in its collection.[14] In memoriam cards became particularly prevalent in Britain after the death of Prince Albert in 1861, when Queen Victoria went into a long period of mourning.[15] Such cards were used to inform people of the death of a loved one and as a way of acknowledging expressions of sympathy received by the bereaved. The cards were sombre in design and usually printed predominantly in black and white. They were often sent in a black-bordered envelope. Many were simply printed with the name and dates of the person who had died, although later cards often included a photograph of them; others included pious verses and images of flowers such as the lily of the valley and the snowdrop.

Victorian Easter greetings card.

Victorian in memoriam cards.

Snowdrops have also featured in memorial sculpture. The 'snowdrop monument' in Lichfield Cathedral in Staffordshire, titled *The Sleeping Children*, is a white Carrara marble sculpture by Francis Chantrey that was first exhibited at the Royal Academy in 1816. It depicts the two daughters of Ellen-Jane Robinson, the widow of clergyman Reverend William Robinson, who died of tuberculosis

Francis Chantrey's *The Sleeping Children* monument in Lichfield Cathedral, 1817, marble; the younger sister holds a bunch of snowdrops.

in 1812, soon after becoming a prebendary of the cathedral. The girls are sculpted lying asleep in each other's arms, with the younger sister holding a bunch of snowdrops.[16] The sculpture was commissioned by Ellen-Jane to commemorate her daughters' tragic deaths. The elder daughter, named after her mother, died in 1813 of burns received when her nightdress caught fire; her younger sister, Marianne, became ill and died on a trip to London the following year.

The sculpture was actually carved by Chantrey's assistant F. A. Lege, and it was said to have been he who had suggested the inclusion of the snowdrops, which grew wild in a nearby wood. The author Edna Jackson, however, records that the idea came from the Scottish poet Allan Cunningham, who had trained as a stonemason and worked as clerk of the works in Chantrey's studio.[17] The sculpture attracted a great deal of attention and admiration. In 1876 the English poet and novelist Jean Ingelow wrote the poem 'The Snowdrop Monument (in Lichfield Cathedral)', which dwells on maternal grief. The priest and poet William Lisle Bowles was also inspired by the work, although in his poem 'Chantrey's Sleeping Children' of 1826 the snowdrops have been transformed into a lily.

The wearing of a snowdrop flower was thought in Victorian times to ensure purity of thought.[18] Snowdrops could also be used as a symbol of innocence and chastity and were carried as an emblem of virginity to warn away unwanted suitors. Richard Mabey reports a tradition by which it was said that a few flowers kept in an envelope served a similar purpose.[19] This idea developed into the Snowdrop Bands of the nineteenth century, which began in Sheffield in South Yorkshire; these groups were devoted to encouraging working-class girls over the age of eleven to lead chaste lives. One such group published a journal called *The Snowdrop*, which carried moralizing stories about young women who had resisted sexual temptation. Members carried cards that pictured a snowdrop as an emblem of purity and chastity, printed with a promise to avoid lewd conversations and immoral literature. The groups also held 'brown suppers' in autumn, during which they potted up snowdrop bulbs; they would meet again in February for 'white suppers' when the bulbs were flowering.[20]

Another journal sharing the snowdrop theme was that produced by the writer sisters Eliza Lanesford Cushing (1794–1886) and Harriet Vining Cheney (1796–1889). Born in Massachusetts, both sisters married Canadians and emigrated to Montreal. Following the deaths of their husbands the two women founded *The Snow-Drop*, a monthly magazine aimed at children aged between six and twelve. It ran from 1847 to 1852 and was chiefly concerned with the social roles and domestic responsibilities considered appropriate for young women at that time. It was modelled on the English children's publication *Peter Parley's Magazine*, which was founded and edited by William Martin and ran from 1839 to 1863. *The Snow-Drop* assumed that children would learn Christian values by example.[21] It contained many short stories and poems with moralizing messages aiming to instil piety and good behaviour in its young readers, such as this admonition from 'Cousin Lizzie':

Be you to others kind and true,
As you'd have others be to you;
And neither do nor say to them,
Whate'er you would not take again.

The image of innocence engendered by the snowdrop has led to its use as a symbol by several children's charities. The Sussex Snowdrop Trust, based in West Sussex, raises money to help provide care for children with life-threatening illness and support for their families. Another British charity, Snowdrop, provides programmes of neurodevelopmental stimulation for children who have brain injuries or conditions such as cerebral palsy and autism.

St Richard's Hospice was established in Worcestershire in 1984 and takes its name from Richard of Chichester, or Richard de la Wyche, who was born in Droitwich in 1197 and was declared a saint in 1262. The hospice logo of two snowdrop flowers is used as a symbol of consolation and was inspired by the lines 'I am come to calm your fears; to console you in the absence of bright days and to reassure you of their return' by the Rev. John Keble. Keble was a leader of the Oxford Movement of High Church Anglicans and the author of the popular work *The Christian Year: A Volume of Thoughts in Verse for the Sundays and Holidays throughout the Year* (1827); Keble College of the University of Oxford was founded in his memory. Patrons of St Richard's Hospice organize a number of money-raising events including a Snowdrop Ball, a dinner dance.

The Dunblane school massacre occurred on 13 March 1996 when Thomas Hamilton entered the school armed with four handguns and killed sixteen children and their class teacher. The Snowdrop Campaign, a pressure group set up in response to the shooting, was founded by friends of the bereaved and was named after the flower, which is usually still in bloom in March in Scotland. It aimed to bring about a total ban on the private ownership of handguns in the UK. Their petition was signed by more than 700,000 supporters and was taken into consideration at the public inquiry into the event chaired by the Hon. Lord Cullen.[22] The snowdrop G. *plicatus* 'Sophie North' was growing in the garden of a Dunblane house bought by snowdrop enthusiast Dr Evelyn Stevens. The flower was named in memory of one of the five-year-old children killed in the massacre. It is popular among exhibitors at the Alpine Garden Society shows, since its broad leaves and large, strongly marked flowers show up well as a pot plant.[23]

The Snowdrop Project is a UK initiative that provides extended support to the victims of human trafficking as they start to build new lives. Once a person is given the right to remain in the UK they may apply for council housing, but the properties are usually unfurnished and often in a poor state of repair. The Snowdrop Project organizes small teams of volunteers to work alongside the survivor of trafficking to paint, decorate and furnish the house or flat, helping to turn it into a welcoming home.

The Perce-Neige association in France raises awareness of and cares for children with learning difficulties. It was started in 1966 by Lino Ventura, an Italian actor who worked in France and himself had a daughter with special needs. He wrote the slogan 'Leur bonheur est notre récompense' (Their happiness is our reward). The charity supports a number of care homes offering different levels of care for those with varying needs.

The charity Child Bereavement UK offers support to families when a child of any age dies or is dying, or when a child is facing bereavement. It has helped to organize the planting of snowdrops in memory of lost loved ones, at several locations, including Sizergh Castle in Cumbria. The snowdrop walk at Sizergh is designed to provide families with a physical remembrance at a place they can visit regularly. Goalkeeper David Martin of the Milton Keynes Dons football team helped the local MK Bereaved Parents Group unveil their snowdrop walk at Woburn Abbey in 2014.[24]

Many gardens with large collections of snowdrops have charity open days in snowdrop season to raise money for good causes. The Snowdrop Charity Day at Walsingham Abbey in Norfolk in 2013 was attended by over 800 visitors, raising more than £3,000 for

A *mărțișor* made from twists of red and white thread, as given to friends in parts of Romania to celebrate the spring.

the Gurkha Welfare Trust and Help for Heroes charities. The UK National Gardens Scheme, founded in 1927 to raise money for the nurses of the Queen's Nursing Institute, is the best-known open gardens scheme and currently gives more than £2.5 million each year to various charities. The scheme produces a 'yellow book' which lists some 3,700 gardens that open, including many with special snowdrop openings that take place before the main garden visiting season begins.[25]

The snowdrop can be used not just as a memorial but as an emblem of new life, as seen in the short story 'Snowdrops' (1996) by the Welsh writer Leslie Norris. In this tale a boy's teacher is showing her class some snowdrops while the funeral procession of a young man who had loved her passes the school. The snowdrop is also used emblematically in other cultures. In Tajikistan, it is believed that the first child to find a snowdrop emerging from the melting snows will have good luck. During the country's annual snowdrop festival the village children pick snowdrops to give to their mothers and other women of the village as a symbol of resurgent life, youth and beauty. The women touch their eyes with the flowers to give thanks that they have lived to see the spring. The children are then given sweets and cakes, a tradition known as *guldardoni*, which is followed by the cooking of a traditional pilaf called *oshi boychechak*, named after the Tajik word for snowdrop, *boychechak*. In Uzbekistan the *boychechak* festival involves groups of children decorating branches of sycamore trees with snowdrops and other flowers and walking around the village handing out flowers and singing songs.[26]

In Romanian legend the beautiful lady Spring had to fight the winter witch who was reluctant to leave Spring's domain. Spring was injured in the fight and drops of her blood melted the snow and brought forth snowdrops. The *Mărțișor* (Little March) festival celebrated on 1 March in Moldova and Romania has an associated legend: in spring, the sun took the form of a human to visit the earth but was captured by a dragon and imprisoned in a dungeon. The world became dark and birds stopped singing. A young man resolved to free the sun and travelled for three seasons to find the dungeon. He fought the dragon and freed the sun but died from his wounds. Where his blood fell, snowdrops sprouted from the earth. The story is remembered with *mărțișors*, which are tassels of twisted white and red wool or silk that are given by men to their loved ones as a symbol of love and to welcome the spring. The *mărțișor* often has a small trinket or coin tied up in the twists of the wool. They were traditionally worn pinned to the lapel until the first fruit trees began to blossom, at which point they were hung up in the branches of a favourite tree.[27]

three

Art and Images

Lisbeth held her fast, while she was taking off her bonnet, and looked at her face as one looks into a newly gathered snowdrop, to renew the old impressions of purity and gentleness.

GEORGE ELIOT, *Adam Bede* (1859)

Flowers by their nature are ephemeral creations, and one of the pleasures of flower paintings is that they can be enjoyed again and again at any time of the year. Mirabel Osler, the English writer and garden designer, has said that 'Portraits of flowers can sustain a gardener through the winter with more propulsion than any nurseryman's catalogue.'[1] That sustenance, which fulfils the plant lover's need for beauty, can be in the form of realistic images of almost photographic detail or those that give just the impression of plants living in an airy environment. Roman still-life frescoes showing fruits and flowers as realistic as any *trompe l'oeil* image survive at Pompeii, having been preserved under mud and ash after the eruption of Mount Vesuvius in AD 79.[2]

Snowdrops are used in art both for their simple beauty and as symbols to express in pictorial form a range of ideas and emotions. They do not appear regularly in medieval art, although the related spring snowflake (*Leucojum vernum*) can be seen in representations of the Virgin Mary by several German painters of the fifteenth century, including the famous *Paradiesgärtlein* (Garden of Paradise) panel painting created around 1410 that is on display at the Städel Museum

Upper Rhenish Master, *Paradiesgärtlein* (Garden of Paradise), *c.* 1410, mixed media on panel.

in Frankfurt, Germany. The artist's identity is unknown and he is usually referred to as the Upper Rhenish Master. It is a charming painting that shows the Virgin Mary reading a book in her garden. She is accompanied by several saints, including St Dorothy, who is depicted picking cherries, and St Barbara, who scoops water from a trough. A small and rather cute dragon lies dead before St George, while the Archangel Michael is accompanied by a little black demon. In early Christian art plants are often depicted in a stylized way, making it difficult to say with certainty what they are modelled on. However, in this painting the flowering lawn and the borders of the garden are filled with readily identifiable plants used as Marian symbols, including the spring snowflake, lily of the valley, iris and peony. The *Madonna of the Strawberries* of around 1420, now on display in the Kunstmuseum Solothurn in Switzerland, was painted by the same artist. It uses fewer types of plants but also includes snowflakes.

Distinctions can be made between botanical illustrations and botanical art or floral still-life paintings. In a pure botanical illustration the emphasis is on giving information with scientific accuracy. It should be possible to distinguish with precision between different species or cultivars.[3] Botanical illustrators generally use watercolours or pen and ink, and although their work may show its subject with almost photographic detail, the style and composition of the picture is usually recognizably that of a particular artist.[4] Such work was frequently commissioned by academic institutions or wealthy plant collectors. In floral art, such as the many botanical still-life paintings produced during the Dutch Golden Age or in the more recent work of painters such as Claude Monet or Georgia O'Keeffe, the plants may again be readily identifiable but the primary aim is aesthetic or symbolic.

Paintings, Drawings and Illustration

One of the earliest botanical illustrations of snowdrops is the beautiful *Snowdrops with a Lady Butterfly* (*c.* 1575) by Jacques Le Moyne de Morgues, owned by the Victoria & Albert Museum. It shows three views of a *Galanthus nivalis* flower and a single painted lady butterfly (*Vanessa cardui*), painted with great attention to detail. Le Moyne was a French artist who travelled to Florida in 1564 with an expedition led by the French Huguenot explorer René de Laudonnière. Le Moyne charted the coastline and the lives of the native Timuca people, including their method of killing an alligator by ramming a pointed log down its throat and flipping it over so they could attack the animal's soft underbelly. Members of the expedition established Fort Caroline (now in Jacksonville) on the St Johns river. This was invaded in 1565 by the Spaniards and most of Le Moyne's drawings were lost when the fort burned. Le Moyne managed to escape and with other survivors returned to France, where he redrew his pictures from memory.[5] He later settled in London where he began working for the explorer Sir Walter Raleigh.

The drawing *Galanthus nivalis: Snowdrops with a Blue-bottle* (*c.* 1584), now at the Fitzwilliam Museum in Cambridge, also shows three views of a snowdrop flower but here with a portrait of a bluebottle fly (*Calliphora vomitoria*). It is the work of the French artist Antoine du Pinet, from Noroy in the Picardy region. It was part of an album containing a dedicatory sonnet and 48 drawings that was produced to celebrate the marriage of Louise of Lorraine, who was Queen Consort of France from 1575 to 1589. After her husband Henri III was assassinated she became known as the White Queen because she spent the rest of her life in mourning clothes, which in France at that time were traditionally white.[6]

The *Hortus Eystettensis* was published in 1613 by Basilius Besler at Eichstätt, near Nuremberg. The title translates to 'The Garden at Eichstätt' and the book is a record of the plants grown in the garden of Besler's patron, Johann Conrad von Gemmingen, the Prince Bishop

Jacques Le Moyne de Morgues, *Snowdrops with a Lady Butterfly*, c. 1575, watercolour and bodycolour on paper.

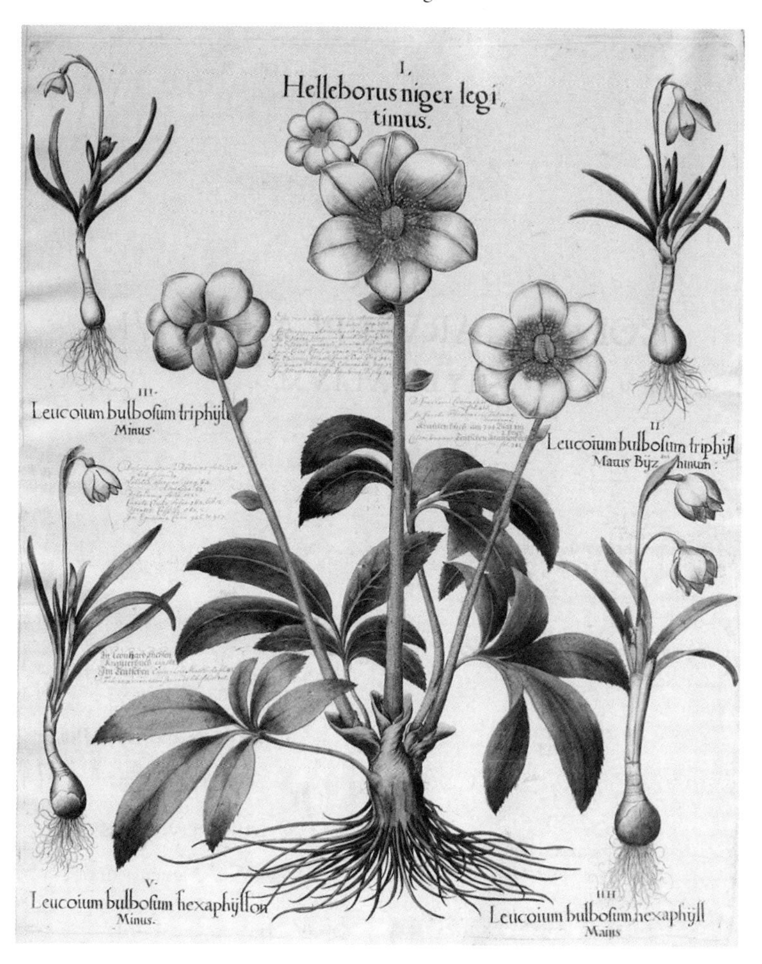

Basilius Besler, '*Helleborus niger*', from *Hortus Eystettensis* (1613).

of Eichstätt. Besler, an apothecary and botanist, worked on the drawings for the 374 copper engravings over a period of sixteen years using a team of artists and engravers. A wide range of flowers, herbs and vegetables were depicted, including plants newly discovered in the Americas such as tobacco and types of peppers (*Capsicum*). The basic edition was in black and white with descriptive text and was intended as a reference book for apothecaries; a hugely expensive luxury edition of the work was produced that had individually hand-coloured plates. The copy held by the British Library was painted by Georg Mack, who took over a year (from 1614 to 1615) to complete the work.[7] One of the plates illustrates a hellebore surrounded by two leucojums and two snowdrops, described in those pre-Linnaean times as *Helleborus niger legitimus, Leucoium bulbosum triphyl Maius Byzanthinum, Leucoium bulbosum triphyl Minus, Leucoium bulbosum nexaphyll Maius* and *Leucoium bulbosum hexaphyllon Minus*.

Still-life paintings were popular particularly in the newly formed Dutch Republic of the late sixteenth and early seventeenth centuries, and though apparently simple often had a deeper allegorical message. Expensive jewellery, gold and silver cups or tankards were used to suggest the vanity of treasuring earthly possessions. The skull has been used since Roman times as a symbol of mortality. In such *vanitas* paintings salvation can often be seen as a chalice amid flowers, while a crucifix in the image aimed to encourage the viewer to reflect on death and resurrection. Butterflies were used to represent transformation and resurrection while the dragonfly stood for transience and ants were included to symbolize an industrious nature. Many common flowers were considered to have symbolic meanings, including the rose, which referenced the Virgin Mary or Venus but also the qualities of transience and love. The tulip could demonstrate nobility and the sunflower divine love and devotion, while the poppy symbolized sleep or death.[8]

Snowdrops feature prominently in many of the paintings of the prolific Flemish artist Jan Brueghel the Elder (1568–1625). Brueghel was born in Brussels to an artistic family. As a young man he travelled to Italy, living first in Naples and then Rome. While there he painted mostly landscapes and history paintings but on his return to Antwerp in 1596 he became better known for his flower paintings and allegorical scenes. He collaborated with Peter Paul Rubens on a number of paintings and Rubens painted a delightful portrait of Brueghel with his family that is now in the Courtauld Gallery in London. In *Basket and Glass Vase of Flowers* (1615) a basket is packed with deep layers of flowers. Snowdrops can be seen in the vase and on the table. A single snowdrop is portrayed in the centre of the bouquet in

Bouquet in a Glass Vase (*c.* 1599), now in Madrid. Other flowers in the bouquet include tulips, irises, narcissi and a scarlet lily with a wonderful sheen to the petals. An amusing addition is the frog squatting on the table beside the vase.

Brueghel's *Flower Garland around the Virgin and Child* (in Milan), painted in 1615, shows a snowdrop at the centre top of the garland. The flowers here, such as scillas, forget-me-nots and chionodoxa, are predominantly blue, to mirror the blue of the Virgin's robe. His painting *Bouquet in a Glass Vase* (1620, in Frankfurt) shows a single snowdrop with what looks like a double hepatica, an angel's tears daffodil (*Narcissus triandrus*), a cyclamen leaf and a blue butterfly.

Snowflakes are sometimes misidentified as snowdrops in descriptions of the works of Brueghel and other botanical still-life paintings,

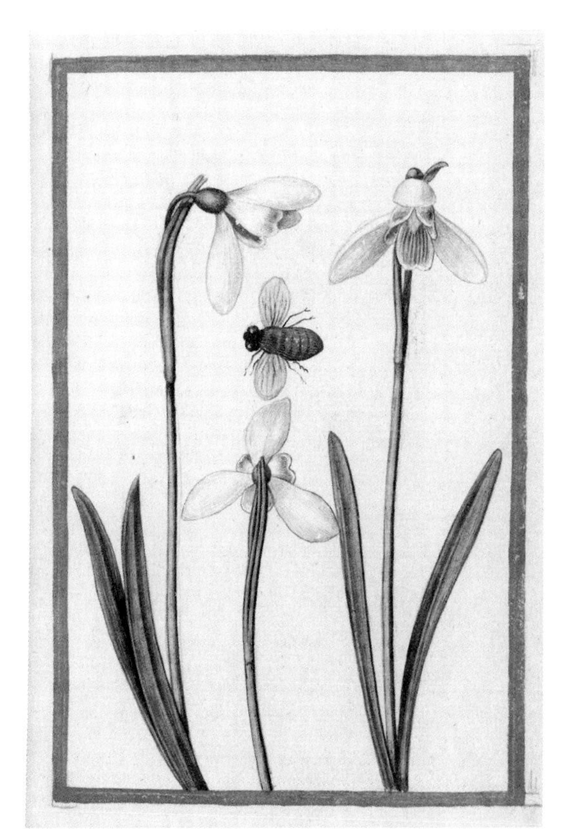

Antoine du Pinet, *Galanthus nivalis: Snowdrops with a Bluebottle*, *c.* 1584.

and this may lead to the impression that the painting was not done from life, for the flowers would not have been in bloom simultaneously. However, the summer snowflake *Leucojum aestivum* does in fact flower with many tulips and, as any gardener knows, flowering seasons can be manipulated by growing plants in particular microclimates in the garden or under glass. The use of greenhouses to enable the wealthy to grow exotic plants developed in the Netherlands at the start of the seventeenth century. They were so-called because they were originally used to house tender 'greens', or evergreens, such as citrus plants, during the winter. It is therefore difficult to say definitively whether a painting was worked on from a live assemblage of plants in one sitting solely by looking at the plants pictured. Paintings could have been worked on in a studio setting over many months, with new flowers added as they came into bloom. This is acknowledged in the painting *Fruit and Flowers in a Terracotta Vase* by Jan van Os in the National Gallery in London. He painted two dates, 1777 and 1778, on the plinth supporting his plethora of flora and fruit.[9]

Artists could of course also have kept sketchpads of flower images to use in assembling their compositions. A charming picture by the popular artist Jan Steen, *Interior with a Painter and his Family* in the Fitzwilliam Museum in Cambridge, shows an artist demonstrating to his son how to sketch a vase of flowers while the mother sharpens his pencils. This oil of around 1670 is a quiet and intimate scene, in contrast to many of Steen's witty but moralizing tales of domestic life.

Most botanical paintings of this period were probably not meant to be taken as representative of a moment of reality, and the deeper, allegorical meaning of a work was considered important. The painting *Flowerpiece* (1663) by the Flemish artist Nicolaes van Verendael, for example, definitely uses flowers that bloom at different times of the year: it shows roses and the late summer-flowering hibiscus combined with spring-flowering narcissi and leucojums. He includes a wilting flower to express the message that beauty is not eternal.

Vertumnus and Pomona by Hendrik van Balen the Elder (*c.* 1575–1632), now in the Musee de l'Hotel Sandelin in Saint-Omer in northern France, depicts a profusion of plants and animals. The story it illustrates concerns the Roman god of the seasons, Vertumnus, who disguised himself as an old woman in order to gain entrance to the orchard of the virtuous nymph Pomona so that he could seduce her. Snowdrops and anemones feature in the foreground along with other plants including tulips, roses, irises, globe artichokes and Madonna lilies. There is a monkey tucking into a melon and a scarlet macaw and peacocks. Balen was a Flemish painter who was born in Antwerp and taught by Adam van Noort, who also taught Rubens; he had many pupils himself, including Anthony van Dyck (1599–1641), who was to become the leading court painter in England.

Festoon of Fruits and Flowers (*c.* 1600) by Jan Davidsz. de Heem depicts a marvellous collection of exotic blooms and fruits hanging from an iron ring by a blue silk ribbon. The fruits include translucent cherries, peaches, a fig, a stripy lemon, a split pomegranate spilling its seeds and a wonderful warty gourd. The flowers are rendered beautifully and are easily identifiable; they include citrus blossom, a white hibiscus, convolvulus and a single snowdrop flower. The painting is enlivened by a number of visiting invertebrates, including a snail and a cabbage white butterfly (*Pieris brassicae*), so realistically painted that it could have just flown in from the garden. De Heem was born in Utrecht but moved to the southern Netherlands and joined the Guild of Saint Luke in Antwerp, after which his work gave way to a more profuse style. He became known for his production of *pronkstilleven*, or paintings in the 'ostentatious still-life' style.[10]

The Belgian painter and botanist Pierre-Joseph Redouté (1759–1840) became known as 'the Raphael of flowers'. His work was very highly regarded and he was very successful despite the changing political events of the time;[11] he was an official court artist of Marie Antoinette and in 1798 Empress Joséphine de Beauharnais, the first wife of Napoleon Bonaparte, became his patron. Redouté painted watercolours of the roses, lilies and other flowers at her residence outside Paris, Château de Malmaison. His work *Les Liliacées*, a collection of stipple engravings in folio size, was produced in eight volumes between 1802 and 1816 from a set of 486 watercolours painted on vellum. It included paintings of not just members of the lily family but those in the Iridaceae (iris) and Amaryllidaceae (amaryllis) families. The common snowdrop, *G. nivalis*, appears in plate 200 in the series.

Robert Thornton, a physician and lecturer in medical botany at Guy's and St Thomas's hospitals in London, decided to produce an English flower book to rival those that Redouté produced for the royal court in France. *The Temple of Flora*, first published in

Jan Brueghel the Elder, *Little Bouquet in a Clay Jar*, c. 1607, oil on panel.

Maria van Oosterwijck, *Vanitas Still-life*, 1668, oil on canvas. One of few female professional artists in the 1600s, van Oosterwijck includes a snowdrop on the book in this painting.

London in 1799, was the third part of Thornton's *A New Illustration of the Sexual System of Carolus von Linnaeus*, a homage to the Swedish scientist. A large number of artists were commissioned to produce the plates, including Abraham Pether, who worked on 28 paintings of flowers. Pether (1756–1812) originally aspired to be a musician but studied painting under the landscape painter and poet George Smith of Chichester; he first exhibited at the Royal Academy in London in 1784 and became known as 'Moonlight' Pether due to the atmospheric lighting of many of his paintings.

Thornton employed Pether to paint the moonlit landscapes for his plates while the flower portraits were done by Philip Reinagle (1749–1833), who was better known for his animal painting.[12] The production expenses of the project caused Thornton huge problems and to raise funds he organized the Royal Botanic Lottery under the patronage of the Prince Regent (George IV). Alan Thomas wrote, 'it is easy to raise one's eyebrows at Thornton's unworldly and injudicious approach to publishing . . . But he produced . . . one of the loveliest books in the world.'[13] A coloured mezzotint of a snowdrop by Pether was engraved by William Ward for *The Temple of Flora*, and for the lottery edition the plate was reworked in aquatint, giving more shading on the snowdrops.[14] The image has the snowdrops with blue and yellow crocus in the foreground, set against a wintry landscape with a snow-covered cottage to the right and a village with a church in the distance.

The Danish artist Johan Laurentz Jensen (1800–1856) specialized in flower painting inspired by seventeenth-century Dutch paintings. He trained at the porcelain factory in Sèvres and later became head artist at the Royal Danish Porcelain Manufactory, now called Royal Copenhagen. He painted arrangements of flowers which often had symbolic as well as decorative value. He included many wild Danish plants as well as exotic imports, which he studied at the Rosenborg Castle gardens in Copenhagen. In 1841 he painted a simple arrangement of white and purple crocus and snowdrops with ivy leaves on a marble ledge.

The French caricaturist Jean Ignace Isidore Gérard was generally known by the pseudonym J. J. Grandville, a name taken from his grandparents, who used it as a stage name for their comedy acts. Grandville came to fame with his work *Les Métamorphoses du jour* (Metamorphoses of the Day, 1829), a series of 73 lithographs of characters with animal heads imposed on the bodies of people satirizing Paris society of the day. *Les Fleurs animées* was published in 83 parts between February 1846 and January 1847 and published posthumously with an English translation in 1847 as *The Flowers Personified.*[15] Aimed at a female readership, the gently satirical text by the journalist Taxile Delord has Primrose telling Snowdrop, 'The spring loves not the winter; youth loves not old age. Thou art at the point of death, and yet thou talkest of love.' Later chapters combine sentimental descriptions of flowers with sound horticultural advice. The snowdrop is described as 'Pensively inclining towards the earth, she seems to regret the obscurity from which she came only to announce the renovation of nature. It is propagated from offsets, which should be detached from the bulbs every two to three years.' It was, however, Grandville's lithographs that were the main selling point of the book. They show flowers personified as elaborately dressed women, often accompanied by anthropomorphized birds and insects. Ciguë (Hemlock) is shown furtively looking over her

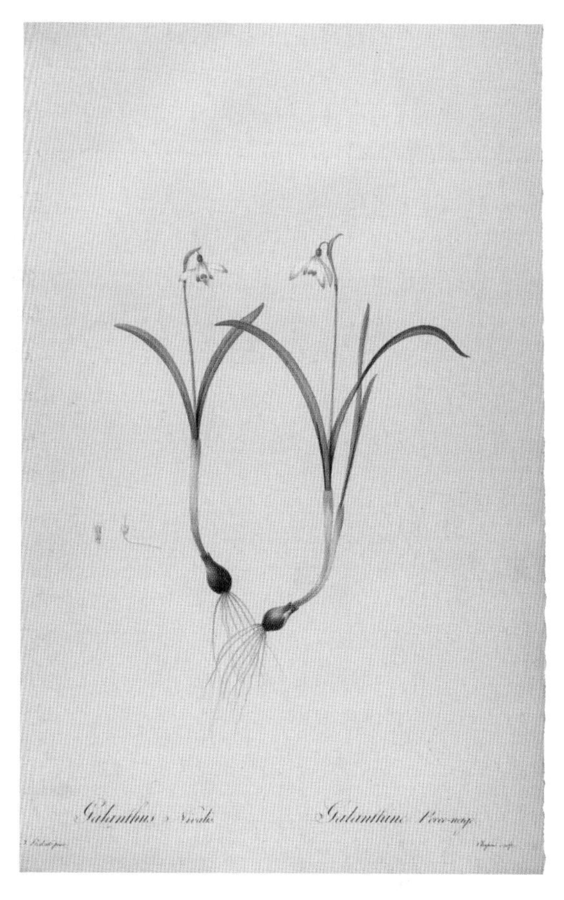

Pierre-Joseph Redouté, *G. nivalis*, 1802, stipple engraving.

shoulder while preparing a poison, with a frog that has already succumbed to a dose dead at her feet. Aubépine (Hawthorn) recoils in fear from a pair of outsized secateurs. Primevère and Perce-neige (Primrose and Snowdrop) are lovely young ladies helping each other through a snowy landscape. Grandville was one of many artists using the new technique of lithography at this time. Lithographs, which had been invented by Alois Senefelder (1771–1834),[16] were much cheaper to produce than etchings and chromolithography was widely used until the introduction of process colour in the 1930s.

Child prodigy John Everett Millais (1829–1896) entered the Royal Academy to study art at the age of eleven. He was a founder member of the Pre-Raphaelite Brotherhood, a group rebelling against the art establishment of the time and taking inspiration from nature and from early Renaissance painting. The inaugural meeting of the Brotherhood took place in 1848 at 7 Gower Street in London, the family home of Millais, who was then only nineteen years old.[17] The other members were painters William Holman Hunt and Dante Gabriel Rossetti, who were sharing a studio in Cleveland Street, and James Collinson. The sculptor and poet Thomas Woolner, artist and critic Frederic George Stephens and Rossetti's brother, William

Abraham Pether, *The Snowdrop*, 1804, colour mezzotint and aquatint.

J. J. Grandville, 'Spring Snowdrop', from *Les Fleurs animées* (1847).

Michael Rossetti, were also included. The first Pre-Raphaelite painting to be exhibited was Rossetti's *The Girlhood of Mary Virgin* (1848–9), with Millais's *Isabella*, based on John Keats's poem 'Isabella; or, The Pot of Basil' (1818), appearing at the Academy a month later in 1849.

In 1851 Millais painted *Mariana*, which was based on the Shakespeare play *Measure for Measure*. Mariana was to be married, but was rejected by her fiancé when her dowry was lost in a shipwreck. In the painting she is shown arching her aching back, having been bent over her embroidery, in a gesture that could also be interpreted as sexual yearning. Millais did not give the picture a title but exhibited it with some lines from Alfred Tennyson's poem 'Mariana':

> She only said, 'My life is dreary –
> He cometh not,' she said;
> She said, 'I am aweary, aweary –
> I would that I were dead!'

The snowdrop depicted in the stained glass on the left of Mariana could be taken to indicate virginal chastity, or to suggest that Mariana is contemplating consolation in death, as is signified by the motto *in coelo quies* (there is rest in heaven).[18]

Millais's drawing *St Agnes's Eve* (1854), based on Tennyson's poem of the same name, shows a nun looking through a window at a snowy convent garden. Snowdrops, the flower of St Agnes, decorate the leaded window of her room and a single flower is tucked into her habit. Millais gave the drawing to Effie Ruskin, wife of his patron John Ruskin, during the time between their falling in love and the annulment of her marriage to Ruskin on the grounds of non-consummation. In a letter to her mother, Effie wrote that the image

> is the most touching thing you ever beheld. The Saint's face looking out on the snow with the mouth opened and dying-looking is exactly like Millais' – which however, fortunately, has not struck John [Ruskin] who said the only part of the picture he didn't like was the face which was ugly . . . I think I see Millais reading the poem to me and talking about it with me.[19]

Gentle Spring is a work by Frederick Sandys (1829–1904) displayed at the Ashmolean Museum in Oxford. In 1857 Sandys had met Rossetti, who was amused by an etching Sandys produced that parodied Millais's picture *Sir Isumbras at the Ford*. Sandys was introduced to other members of the Pre-Raphaelite Brotherhood, who influenced the way he worked. *Gentle Spring* is said to represent the ancient Roman goddess Proserpina, who was abducted by Pluto, the god of the underworld. Proserpina's mother, Ceres, the earth goddess, in her desperation at losing her daughter prevents fruit from growing. Jupiter orders Proserpina's freedom but because she has eaten six pomegranate seeds she must spend six months of the

John Everett Millais, *Mariana*, 1851, oil on wood.

year in the underworld. When Proserpina emerges each spring, Ceres brings forth flowers to celebrate her return.

Sandys, inspired by the Brotherhood's ideals of fidelity to nature, painted *Gentle Spring* in the garden of the poet and novelist George Meredith (1828–1909). It shows a flame-haired Proserpina standing among meadow flowers in an orchard of blossoming trees with a rainbow arching above her. Butterflies and dandelion clocks depict the transience and fragility of life. Her classically draped dress holds a number of accurately portrayed flowers and she has a circlet of snowdrops and crocuses in her hair. The painting is technically superb, although Proserpina looks curiously lacking in joy for a woman just released from six months in the underworld. Unlike Rossetti's own Proserpina paintings, Sandys's does not include the pomegranate fruit that sealed her fate, which does create doubt about whether the subject is in fact Proserpina. The painting was exhibited at the Royal Academy in 1865 and the catalogue included a sonnet by Sandys's friend Charles Algernon Swinburne, in which the figure is compared to the Virgin Mary: 'O virgin mother! of gentle days and nights, / Spring of fresh buds and spring of soft delights.'[20] The model is thought to have been Millie Jones, the sister of Sandys's common-law wife, Mary.

Rossetti himself used snowdrops in both his poetry and his paintings. His sonnet 'True Woman – 1. Herself' from *The House of Life* talks of 'The wave-bowered pearl, – the heart-shaped seal of green, / That flecks the snowdrop underneath the snow.' The title of his snowdrop painting *Blanzifiore* (1873) translates as 'white flower'; it was a popular given name in the Middle Ages and features in medieval literature. 'Blanziflor et Helena', for example, is the title of the final part of Carl Orff's work *Carmina Burana* (1935–6), which was based on 24 poems from a medieval collection. *Blanzifiore* was initially painted as one of a series of images of Proserpina using Jane Burden Morris – wife of William Morris and Rossetti's muse – as the model but the work was unfinished and Rossetti cut the canvas down, making it into a head and neck portrait.[21] Morris is shown

Dante Gabriel Rossetti, *Blanzifiore*, 1873, oil on canvas.

with primroses in her luxurious auburn hair and holding a trio of snowdrops. The snowdrops are fairly neat doubles, possibly selected specimens of *G. nivalis* 'Flore Pleno'; the flowers, unusually, look as though they have four outer petals. Rossetti's painting is now in the private collection of Andrew Lloyd-Webber.

The art of the Pre-Raphaelites was influential to the developing work of the artist and book illustrator Walter Crane (1845–1915). Crane was born in Liverpool and studied the work of the

The Snowdrops first upon
the scene,
White-crested braved King
Frost's demesne:

Walter Crane, illustration from *Flora's Feast: A Masque of Flowers* (1889).

Pre-Raphaelites while apprenticed as a wood engraver to William James Linton. Linton was impressed with Crane's illustration of Tennyson's poem 'The Lady of Shalott', which was exhibited at the Royal Academy in 1862. Crane became a prolific illustrator of children's books but also worked on designs for ceramic tiles, stained glass, wallpaper and textiles. In his design work he moved away from Pre-Raphaelite principles, believing that 'the artist works freest and best without direct reference to nature, and should have learned the forms he makes use of by heart.'[22]

One of Crane's best-known publications was *Flora's Feast: A Masque of Flowers*, first published in 1889. Written and illustrated by Crane with forty colour lithographs, this was a parade of humanized flowers in the tradition of Grandville. The illustrations incorporate Crane's verse, which follows Queen Flora awakening the flowers in her garden as she passes through the seasons. Snowdrops, somewhat unusually, are shown not as demure ladies but as young men with floral helmets and each armed with a sword and spear with which to defeat winter: 'The Snowdrops first upon the scene / White-crested braved King Frost's demesne.' The parade continues

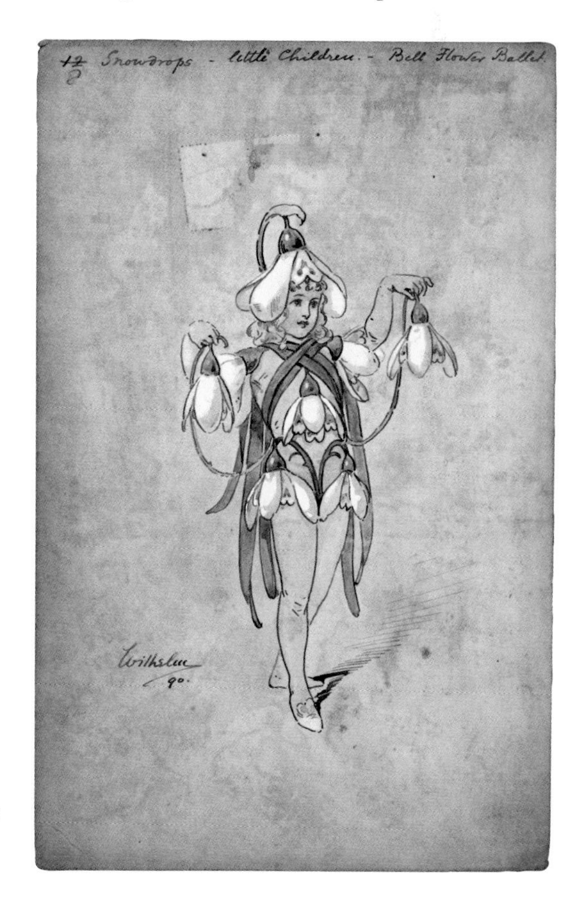

Snowdrop costume for a pantomime by the English costume designer Wilhelm, 1890.

with crocuses and then hyacinths, whose 'bells did ring / To swell the music of the Spring', and on to the late summer sunflowers and the Christmas rose.

Many of the costume designs of Charles William Pitcher (1858–1925) have a similar character to Crane's anthropomorphic flowers. Pitcher, who worked under the name Wilhelm, was an artist and prolific costume designer who was best known for his designs for pantomime costumes for London's Drury Lane Theatre. His design for a snowdrop to feature in the 'Little Children in the Bell Flower Ballet' in the pantomime *Dick Whittington*, performed at Crystal Palace on 24 December 1890, shows a small boy dressed in white decorated with green ribbons and wearing a snowdrop flower hat while holding aloft other snowdrops.

The Swiss artist Eugène Samuel Grasset (1845–1917) was an important contributor to the Art Nouveau movement in France. His work included book illustration and jewellery design, and his posters influenced many other artists including Alphonse Mucha and Maurice-Pillard Verneuil. His painted botanical illustrations, including subjects such as lily of the valley, wisteria and lilac, are very finely worked. Grasset's portfolio of wood engravings depicting women in seasonal costumes was issued as a calendar called *Les Mois* (The Months) by the Paris publisher G. de Malherbe in 1896. The image for February shows a woman training fruit trees in an orchard with snowdrops as the flower for the month. Grasset's book *Plants and their Application to Ornament* (1897), which has 72 colour plates, was a collaboration with several other artists, including Verneuil (1869–1942), who created the designs for Grasset's book. Two of Verneuil's plates show how a botanical drawing of the snowdrop could be developed into a design feature that could be used for fabrics, wallpaper and borders.

The Russian-born artist Elena Luksch-Makowsky (1878–1967) was part of the modernist art movement in Vienna at the beginning of the twentieth century, and she worked closely with the Vienna Secessionist group of painters, headed by Gustav Klimt, who objected

Maurice-Pillard Verneuil's Art Nouveau snowdrop design, 1897.

to the prevailing conservatism of the Association of Austrian Artists. Luksch-Makowsky's *Self-portrait with her Son Peter* (1901) was originally titled *Ver sacrum*, or Sacred Spring, which was the title of the Secession's magazine and had connotations of regeneration.[23] In it she depicts herself as a Madonna-type female in a dark-red robe, merging into the dark background while her son Peter is bathed in gentle light. The painting has echoes of the traditional Russian icons of her heritage. Her son Peter clutches a handful of snowdrops, symbolic of rebirth and resurrection after the death of winter.

One of the most interesting examples of snowdrops in painting features in *The Wounded Angel* (1903) by the Finnish Symbolist painter Hugo Simberg. It illustrates a winged angel with bandaged head being carried on a stretcher by two boys. The first boy is looking at the path ahead but the other regards the viewer with a haunting expression. In the angel's hand is a bunch of snowdrops, here used as symbols of healing and rebirth. The landscape in the

Hugo Simberg, *The Wounded Angel*, 1903, oil on canvas.

Beryl Fowler, *Woman Holding Snowdrops*, 1905, oil on canvas.

background has been identified as Eläintarha Park in Helsinki, with the waters of Töölönlahti Bay clearly visible.[24] The boys carry the injured angel towards what was then known as the Blind Girls' School and Home for Cripples, a charitable institution that was opened in 1865. Simberg (1873–1917) studied art from the age of eighteen, latterly as a private pupil at the wilderness studio of the Symbolist painter Akseli Gallen-Kallela (1865–1931), whose fantasy figures represented angels, the devil and death. Simberg saw work by the Pre-Raphaelites in London and also travelled to Paris and Italy. *The Wounded Angel* was painted while Simberg himself was being treated for a nerve disorder. It is one of the most recognizable of the artist's works and was voted Finland's 'national painting' in a vote held by the Ateneum Art Museum in Helsinki in 2006. Maija Tanninen-Mattila, director of the Ateneum, expressed her surprise and delight at the choice of Simberg's painting, which she considers to be a difficult work open to multiple interpretations.[25] Between 1905 and 1906 Simberg painted a larger version of this work as a fresco in Tampere Cathedral, a Lutheran church in southern Finland.

A quiet snowdrop painting by Beryl Fowler is on display at The Beacon Museum in Whitehaven, West Cumbria. Fowler (1881–1963) and her husband Frank were both students of the German-born artist and pioneering film-maker Hubert von Herkomer (1849–1914). Much of Fowler's work depicts images of rural life in the

Myles Birket Foster, *Snowdrops*, 1868, watercolour on paper.

Eskdale area of Cumbria. The *Woman Holding Snowdrops* was painted in 1905 and depicts a woman in a white linen bonnet gazing contemplatively at a handful of snowdrops. *Gathering Snowdrops* (1906) by Edward Atkinson Hornel uses similar quiet colours to those in Fowler's painting but pictures three lively young girls filling a basket with snowdrops. Hornel (1864–1933), an Australian-born Scottish painter, produced a number of similar paintings of children in gardens and woodland.

Edith Holden (1871–1920) was born in Kings Norton near Birmingham. She worked as an illustrator of children's books, such as *Woodland Whisperings* (1911) by Margaret Rankin. At the age of forty she married a sculptor seven years her junior, Ernest Smith, to the disapproval of her family. The couple moved to Chelsea in London where they met many of the leading artists of the day. Tragically, Holden drowned in the Thames when she fell into the water while trying to reach a bough of chestnut buds to use as a subject for a painting. In the mid-1970s Holden's great-niece Rowena Stott found a collection of Holden's watercolours and nature observations made in 1906 as an exercise in the observation of nature for her art students at Solihull School for Girls, where she taught art on Fridays. The collection was published in 1977 in a facsimile version, titled *The Country Diary of an Edwardian Lady*, and sold more than six million copies over the next thirty years. An image for February shows a robin searching for food through the leaf litter in which snowdrops and aconites are growing.

The British artist and social reformer Winifred Gill (1891–1981) came from a Quaker family in Surrey. Her linocut *Snowdrops* features hand-coloured snowdrops in a pale blue jug. It was produced for the Omega Workshops, a design studio established in 1913 by members of the Bloomsbury Group under the direction of Roger Fry, Duncan Grant and Vanessa Bell.[26] Based in Fitzroy Square, London, the Omega Workshops ran for just six years but were very influential in interior design. They aimed to help young artists earn a living and to build upon the ideas of William Morris in bringing

Cicely Mary Barker, 'The Snowdrop Fairy', from *Flower Fairies of the Winter* (1985).

together the fields of art and industry. The range of products included furniture, pottery, book jackets and dress designs. All work from the workshop, for which Gill did much of the textile design, was marked with the Greek letter omega (Ω) but did not feature the individual artist's name. Clients included George Bernard Shaw and Gertrude Stein.[27]

One of the most popular twentieth-century children's book series was the Flower Fairy books of Cicely Mary Barker. The first title, *Flower Fairies of the Spring*, was published by Blackie in 1923. Barker (1895–1973) suffered from epilepsy and was educated at home by a governess. She had no formal artistic training but as a young woman joined the Croydon Art Society; she was an admirer of the art of the Pre-Raphaelites, particularly Millais and Burne-Jones. Barker's watercolours combined accurate botanical details with lively depictions of children, often using the children who attended her sister's kindergarten as models. She painted plants from life, and if Barker did not have a suitable specimen available, staff at the Royal Botanic Gardens, Kew, would provide her with the plants she required. *The Song of the Snowdrop Fairy*, known for the line 'The Fair Maids of February stand in the snow', was published posthumously in *Flower Fairies of the Winter*. Barker's illustration shows a demure, blonde-haired child fairy standing barefoot in the snow.

Ida Rentoul Outhwaite (1888–1960), an Australian book illustrator, was first published at the age of fifteen when she illustrated a story called 'The Fairies of Fern Gully' written by her sister Annie Billabong for the *New Idea* magazine. Outhwaite worked mostly in pen and ink or watercolour, and her illustrations were very popular in her native Australia because many of them combined the craze for fairies with local wildlife such as kangaroos and koalas. For the book *Bunny and Brownie: The Adventures of George and Wiggle* (1930) Outhwaite produced eight colour plates with more European settings, including 'Up North', in which three snowdrop fairies dance among clumps of snowdrops flowering amid the melting snow. Queen Mary, wife of George V, helped to popularize Outhwaite's work in England by sending postcards featuring her art to friends during the 1920s.[28]

Ceramics

Since the invention of porcelain in Han Dynasty China (206 BC–AD 200),[29] China has had a long tradition of painting flowers on porcelain and such wares were much admired in Europe. The Meissen factory in Germany was established in 1710 after the discovery in 1708 by Ehrenfried Walther von Tschirnhaus of a method to produce a hard, white, translucent porcelain using ingredients such as kaolin clay and alabaster, mined from Colditz.[30] Much early Meissen is painted with Chinese-style flowers and plants, including cherry blossom and peonies.

One of the best-known European pottery firms is that established by Josiah Wedgwood on 1 May 1759 at Burslem in Stoke-on-Trent. Wedgwood's 'Porkoipins for snodrops' [*sic*] was first made around 1775 and was a black basalt 'hedgehog' bulb pot which was used to grow and display snowdrops. Holes in the body of the pot enabled the bulbs to grow through and flower. It was later also made in terracotta or a green earthenware glaze and was available in three sizes. In more recent times Wedgwood has produced a number of snowdrop items, including designs by Susie Cooper, and a modern range of dinnerware and decorative items called Sarah's Garden.

The Royal Danish Porcelain Manufactory, which later became Royal Copenhagen, was founded by the pharmacist Frantz Heinrich Müller. Müller had experimented with hard porcelain made from quartz, kaolin and feldspar and in 1774 started looking for potential investors for his Danish porcelain factory. The Queen Dowager Juliane Marie and her son Frederik became partners in the company the following year. The Danish royal family remained involved in the factory for almost a century until it passed into private hands in 1868. In 1882 the factory was amalgamated with the Aluminia faience factory and moved from Copenhagen to new premises in Frederiksberg.

One of Royal Copenhagen's oldest and most famous lines is the Flora Danica service in which, unusually, the floral images were not chosen primarily for aesthetic reasons. In tune with the spirit of the Age of Enlightenment, the cultural movement characterized by a strong belief in science and rationality, it was decided to use exact copies of the botanical images from the highly regarded work *Flora danica*.[31] This encyclopaedia of the national flora of Denmark included 3,240 hand-coloured folio engravings that were produced between 1761 and 1883. It was the idea of Georg Christian Oeder, who was appointed professor of botany by royal decree in 1752. He wanted to create a national flora of all the wild plants in Denmark to widen appreciation of botany and enhance the knowledge of the medicinal and other properties of plants. There was a hand-painted edition and a cheaper print version, and to ensure that the work became widely used free copies of the plain edition were given to bishops, who distributed them to clergymen and schools. The project was supported financially by the Crown.

The original Flora Danica dinner service, which features *Galanthus nivalis*, was commissioned by Crown Prince Frederik, later King Frederik VI, in 1790 as a gift to Catherine II of Russia, although unfortunately she died while it was still in production. The royal librarian in Copenhagen sent the drawings to the porcelain works, where Johann Christoph Bayer painted the images onto the service. Bayer (1738–1812) was born in Nuremberg. His father owned a glass-painting factory but he was more interested in porcelain and travelled to Copenhagen to work for the royal manufactory. Items in the service ranged from egg cups to vast tureens; there were some problems transferring designs from the rectangular format of the original images to the round or oval shapes of the service, and sometimes compromises had to be made. Many pieces had serrated edges, hand-cut perforations and finely modelled flowers on lids, covers and handles.

The Flora Danica service was first used on 29 January 1803, for the king's birthday, and has continued to be used on state occasions. Of the original 1,802 pieces delivered, 1,530 still survive and can be seen at the Rosenborg Castle museum. Production was resumed in 1862 when a Flora Danica dinner set was made as a wedding present for the Danish princess Alexandra on her marriage to the future King Edward VII of England. This service is held by the Royal Collections Trust and kept at Windsor Castle.[32]

Royal Copenhagen continues to manufacture the hand-painted Flora Danica porcelain today. In 1959 Royal Copenhagen produced a less costly snowdrop design, which was a charming oval plate hand-painted with a view of snowdrops blooming through the snow. The cold, grey landscape emphasizes the resilience of the flowers to the wintry weather.

In Sweden the flower painter Gunnar Wennerberg (1863–1914) designed simple repeating motifs of stylized flowers, including snowdrops and lily of the valley, to decorate the fine bone china of the Gustavsberg Porcelain Factory. The snowdrop design was used for the General Art and Industrial Exposition of Stockholm in 1897.[33] Wennerberg trained at the Sèvres factory in France and designed glassware for Kosta from 1898 to 1909. He was brought in to the company in response to criticism at the Stockholm exhibition that Kosta products were too similar to those produced in other parts of Europe.

The Hopkins Brothers of Lambeth in London were the third generation of a family of potters. Alfred George Hopkins (1884–1940) was established as a studio potter by 1915 and had one-person exhibitions at the Fine Art Society (1927–8) but also collaborated with his brother, Henry Loveday Hopkins (1885–1951). Together they created many plates, vases and figurines, particularly in stoneware.[34] Both brothers held classes in pottery at Camberwell School of Art. A Hopkins earthenware dish made in 1916, held in the Victoria & Albert Museum, shows the influence of the Art Nouveau movement on the brothers' work. The design, in green and white on

Galanthus nivalis, from the Danish botanical work *Flora danica* (1761–1883).

a black background, is of a young nude woman standing among the swirling stems of a large flowering lily. At her feet are three flowering snowdrop bulbs.

The English art pottery firm Moorcroft began in 1897 as a studio of the large ceramics company James Macintyre & Co. In 1913 William Moorcroft, who designed for Macintyre, moved with his team to a new factory in Sandbach Road in Burslem, Stoke-on-Trent, funded by the famous London department store Liberty.[35] Moorcroft has used snowdrops in several of its designs. For example, in 2011 Nicola Slaney created an alpine meadow design featuring snowdrops with fritillaries, which decorated miniature vases. In 1999 Lise B. Moorcroft, the granddaughter of William Moorcroft, produced a retro-style design of white snowdrops on black with orange highlights and snowdrop vases using a slip trailing technique called tube lining. The design was first sketched freehand in pencil onto the pot and the colours were built up over several firings. Moorcroft Enamels, which ran from 1998 to 2006, also used snowdrops in a design of the white flowers on a blue background. Another snowdrop design was produced in 1999 exclusively for the members of the Moorcroft Collectors Club.

The Portmeirion pottery was founded in 1960 by pottery designer Susan Williams-Ellis and her husband, Euan Cooper-Willis. Williams-Ellis had been designing for the company A. E. Gray, which sold gift items at a shop in Portmeirion village in north Wales. In the early 1970s she visited an antiquarian bookseller in London and purchased *The Universal Herbal* (1817) by Thomas Green. The detailed botanical illustrations in this book inspired her to create a new range of pottery that used images from this and other botanical works. The range was named Botanic Garden after an eighteenth-century poem, 'The Botanic Garden' by Erasmus Darwin. Botanic Garden was launched in 1972 and included 32 motifs in the first year. The crocus and snowdrop design was first introduced in 1980 and was used for twenty years. It was based on an image in the book *The Romance of Nature* (1836) by Louisa Twamley.

Twamley (1812–1895) was born in Birmingham and published her first volume of poems at the age of 23. She married her cousin, Charles Meredith, and went with him to live first in New South Wales, Australia, and then in Tasmania. She wrote a number of books, many of which she illustrated herself, but struggled financially when the bank where her savings were held collapsed. *The Romance of Nature*, subtitled *The Flower-seasons Illustrated*, began with the snowdrops and crocus and travelled through a year of flowers. It was illustrated with colour plates and descriptions interwoven with Twamley's own verses and those of other poets, quotes from famous authors and heartfelt, if somewhat sentimental, suggestions, such as: 'if people would be but wise enough through life to derive enjoyments from such innocent pleasures as delighted them in childhood, we should find far fewer sour tempers, cold hearts and narrow minds in the world.'

Les Femmes de la Révolution is a modern installation of porcelain and cherrywood produced in 2003 by the highly regarded Scottish poet, artist and gardener Ian Hamilton Finlay (1925–2006). On display at Aberdeen Art Gallery, the work consists of a beautiful table laid for a dinner party, with each plate painted with a wild flower and the name of a woman who played a part in the French Revolution. The opposing sides, republicans and monarchists, are imagined dining together so that Marie-Antoinette sits opposite Jean-Paul Marat's assassin, Charlotte Corday. The plate for Manon Roland, or Madame Roland, known for her cry 'O Liberty, what crimes are committed in thy name!', is painted with a pure, unblemished snowdrop. Roland was executed by guillotine for her republican ideas. Most of the imagined guests at the table suffered violent deaths and so the dinner party becomes in effect a last supper. Little Sparta, Finlay's garden at Dunsyre in the Pentland Hills near Edinburgh, frequently uses revolutionary iconography to represent the struggle between nature and order. In 2004 Little Sparta was voted in a poll of art historians, artists, museum directors and other art professionals to be the most significant Scottish work of art of all time.[36]

Ceramic tiles for walls and floors may be designed with the same care that artists use for more obviously decorative items. There was a boom in the use of tiles in public and domestic buildings in the latter half of the nineteenth century. Tiles have many properties that favour their use over other materials: they are hard-wearing, fire-resistant and easy to clean. The firm of Minton & Co. dominated the British tile market at that time and examples of their floor and wall tiles were shown at the 1851 Great Exhibition in London.[37] The firm split in 1868 when Colin Minton Campbell took charge of what became known as Mintons, producing china and pottery at his China Works amid a dispute regarding the rights to use the Minton name. He continued to produce wall tiles and employed notable designers including Christopher Dresser.

Dresser (1834–1904) was born in Glasgow and attended the Government School of Design (now the Royal College of Art) in London from the age of thirteen. He had a great interest in plant life and was appointed Professor of Artistic Botany at the Department of Science and Art, South Kensington, in 1855. He started selling his first designs in 1858 and published a number of works, including three textbooks on botany. His work *Principles of Decorative Design* (1873) was an important resource for members of the Arts and Crafts movement, which began in Britain around 1880.[38] Dresser was regarded as a pivotal figure in the Aesthetic movement. One of his designs for Campbell was used around 1875 on tiles made by the 'dust pressing' method, which involved compressing nearly dry clay between two metal dies. Dresser's design shows stylized snowdrops set against a black background within a border. The upper border has a repeated floral image that suggests the view inside the bell of a snowdrop, while the lower border consists of a repeating image of bulbs.

In around the same year a set of twelve tiles was produced at Mintons China Works that was known as the Spirits of the Flowers series.[39] Designed by the Scottish architect and designer John Moyr Smith, each featured a naked fairy emerging from different flowers including snowdrops, lily of the valley, foxgloves and water lilies. Moyr Smith (1839–1912) produced many similar series for Mintons, including designs of water nymphs, nursery rhymes and scenes from *The Pilgrim's Progress*, and also designed for the Leeds-based firm Burmantofts. He produced work for Christopher Dresser and supplied designs to the Arthur Silver Studio for textiles and wallpapers. Moyr Smith's first book, *Studies for Pictures: A Medley* (1868), was dedicated to Dresser.

Glass

Coloured glass has been produced since ancient times and stained glass windows are known to have been used in churches since at least the seventh century. The earliest surviving coloured glass window in England, at the twin monasteries at Wearmouth-Jarrow in Northumberland, dates from the ninth century.[40] Perhaps surprisingly, considering their close link with the Church and the Candlemas festival, snowdrops are not often seen in the older stained glass windows of British churches. St Ethelbert's church at

Snowdrop thimble collection, including ceramic and glass examples.

Herringswell in Suffolk has a series of windows that are mostly the work of the renowned stained glass artist Christopher Whall and his pupils. The newest was installed in the south transept in 1992 by the artist Dean Cullum. It was donated by Ivona Mayes-Smith and illustrates the theme 'Herringswell in spring', depicting snowdrops and daffodils as well as Tiffany, the spaniel that belonged to Mayes-Smith.

The parish church of St Peter at Benington in Hertfordshire has two stained glass panels in the porch's east window. One, decorated with daffodils, commemorates the rector William Eustace Mills (1881–1957). The other, with a circle of snowdrops, is in memory of his wife, the long-lived Everilda Louise Tindall Mills (1889–1992). The gardens of Benington Lordship next to the church are noted for their snowdrops. The gardens surround the Georgian manor house and remains of a Norman castle and moat, and swathes of snowdrops fill the moat each year in early spring. There is also an interesting and varied collection of different snowdrop cultivars in other parts of the garden.

Avendale Old Parish church at Strathaven in South Lanarkshire, Scotland, had a new stained glass window dedicated in 1996. Commemorating the parishioner Isabel Simpson, it features twelve flowers to represent the Apostles, including three primroses for the Holy Trinity, four snowdrops for the four Evangelists and five angels' tears to trumpet the resurrection of the faithful. The window is the work of Crear McCartney, who attended Glasgow School of Art in the 1950s. The Anglican church of All Saints in Denmead, Hampshire, has a tall lancet window on the west side that celebrates the natural world, featuring animals such as a squirrel and a hare, bees and blue tits and a spider on a well. Flowers represented include foxgloves, bluebells and snowdrops. The snowdrops are acid-etched on a blue-flashed glass. The window was created by the stained glass artist Jude Tarrant and installed in 2013.

In the fantasy film *Stardust* (2007), based on Neil Gaiman's novel, the hero Tristran Thorn meets an enslaved princess, who offers him a glass snowdrop in exchange for a kiss. Indeed, snowdrops fashioned from glass are available from several modern manufacturers. Snowdrops have featured as a decorative element on vases and other glassware fairly regularly through the years. The ancient technique of creating cameo glass, in which the glassmaker carves through fused layers of different coloured glass, underwent a revival in the mid-nineteenth century as a response to the display from 1810 onwards of the Portland Vase in the British Museum in London. This Illyrian vase, thought to date to between AD 1 and 25, is made of deep blue glass with a overlying white glass cameo featuring mythological characters and a large snake.

Emile Gallé was one of the best-known designers of cameo glass. He was considered one of the leaders of the French Art Nouveau movement during the late nineteenth century. Gallé (1846–1904) studied art, botany and philosophy but went to work in his father's glass factory in Nancy after the Franco-Prussian War of 1870. He exhibited at the Paris Exhibition of 1878, where his work attracted considerable attention. Gallé's 'Vases de Tristesse', or 'sadness vases', were produced in the latter years of the nineteenth century. The term was used by Gallé to describe those vases whose mood evoked feelings of sadness or melancholy thoughts. Several were designed in memory of dead friends and often have appropriate words or names engraved on them. They also frequently feature dead or dying flowers or flowers such as the snowdrop which can symbolize the brevity of life.[41]

The Daum brothers, Auguste (1853–1909) and Antonin (1864–1930), started work for their father, Jean, who acquired a glass factory by accident when as a notary he lent money to the proprietors, who were subsequently unable to pay their debts. The factory in Nancy in the northeast of France started by producing drinking glasses but in 1891 developed a range of decorative glass inspired by the work of Gallé. A typical example of the products created is an etched and carved cameo glass snowdrop vase made around 1900, which featured carved snowdrops on a blue mottled background.

The Daum studio revived the ancient Egyptian method of glass casting known as *pâte de verre*, or glass paste. It uses the lost wax casting method, which is also used for other art objects from bronze sculpture to metal jewellery. Glass cases made using this technique include one with black snowdrops and muscari on a white ground which was created around 1910.[42]

The Muller family of ten siblings – nine boys and a girl – fled to the town of Lunéville near Nancy during the Franco-Prussian War. Désiré and Eugène, the two eldest brothers, started working at the Gallé factory in 1885. Henri Muller started his own decorating workshop and the rest of the family later joined the firm, making cameo and other glass from about 1895 to 1933. They used a combination of acid-etching and wheel carving, producing often very complex vases made up of as many as nine layers of coloured glass. Snowdrops feature on some of their creations, including a squat vase decorated with golden snowdrops emerging from snowy ground.

Textiles

Tapestries have been made in Europe for thousands of years, and fragments of a woollen tapestry found preserved in the desert of the Tarim Basin in China have been dated to the third century BC. It is thought to have come from the Graeco-Bactrian Kingdom in Central Asia and had been made into a pair of man's trousers, suggesting it may have been a decorative trophy of war. Tapestries were extremely popular in the Middle Ages as they combined the aesthetic qualities of a painting with the warm, insulating properties of fabric. Many tapestries of this period, such as the famous Lady and the Unicorn series of around 1500, displayed in the Musée de Cluny in Paris, were created in the *millefleur* (thousand flowers) style.[43] Many of the flowers depicted can be individually identified.

In the nineteenth century the writer and designer William Morris (1834–1896) resurrected the art of tapestry making in the medieval style. Morris's company made a large number of tapestries for home and ecclesiastical use designed by the artist Edward Burne-Jones, with much of the floral work in the foregrounds and backgrounds done by John Henry Dearle. The Holy Grail series of tapestries was designed by Burne-Jones, having been commissioned from Morris & Co. by William Knox D'Arcy in 1890 for his dining room at Stanmore Hall in Middlesex. 'The Arming and Departure of the Knights of the Round Table on the Quest of the Holy Grail' is now in the private collection of Jimmy Page, the guitarist and founder of Led Zeppelin.[44] The flora, depicted in the *millefleur* style, may not all be entirely botanically correct but the tapestry certainly includes parrot tulips and what appear to be snowdrops at the feet of one of the maidens.

The Orchard tapestry, now in the Victoria & Albert Museum in London, was Morris's own design for a tapestry, created in 1890. It shows an orchard of fruit trees, including apples, pears and olives, as a backdrop to a row of figures in medieval-style dress. The figures hold up a banner with a poem celebrating the rhythm of the seasons. The flowers in the foreground include rather stylized pansies, poppies, fritillaries and bellflowers. A small clump of snowdrops grows at the feet of the first figure.

The village of Ufford in Suffolk produced a large tapestry to celebrate the turn of the new millennium. The tapestry was worked on by many members of the community and portrays the life and environment of Ufford in the year 2000. The design shows the village of Ufford within its parish boundary and features prominent buildings, leisure pursuits and other associations and symbols of the area, such as the Suffolk Punch horse breed. The four corners of the tapestry show the seasons and the flora and fauna to be found in Ufford, including winter plants such as aconites, mistletoe and snowdrops depicted with the robin, moorhen and stoat. The finished work is 1.2 metres by 0.9 metres (4 by 3 feet) and contains nearly 250,000 stitches, which took the six principal tapisers 21 weeks to create.[45]

The textile designer Raymond Honeyman is one of the main designers for the modern company Ehrman Tapestry, which makes needlepoint kits. Each design consists of approximately 50,000 stitches and takes several weeks for the designer to paint. Honeyman studied at art college in Dundee; his first professional design, *Snowdrops*, was bought by the famed London department store Liberty. The complex repeating patterns of the Arts and Crafts period were a source of inspiration. A kit of the design 'Snowdrops' was made by Ehrman. Liberty has used snowdrops as an inspiration in other designs, including its popular fabric ranges. Its design 'Martina', on tana lawn, a lightweight cotton fabric using fine, high-count yarns which results in a silky feel that can be used for dressmaking, is decorated with a repeated pattern of snowdrops and crocuses.

In 2010 the Fashion and Textile Museum in Bermondsey, south London, held an exhibition devoted to the Lancashire fashion company Horrockses, which was established in 1946. They chiefly produced women's dresses and beachwear. Although the dresses were mass-produced the firm was renowned for using good-quality cottons and top fashion and textile designers and attracted clients including Princess Margaret and Queen Elizabeth II. One dress featured in the exhibition was the snowdrop pattern of around 1949, which used fabric designed by the artist Graham Vivian Sutherland. Sutherland

Bobbins handpainted with snowdrop designs and bobbin lace.

(1903–1980) trained at Goldsmith's College art school and was an official war artist during the Second World War. After the war Sutherland designed the tapestry of *Christ in Glory* for the new Coventry Cathedral.[46]

Collectors' Items

Britains toy company is best known for its range of lead soldiers, which were made by a process of hollow casting that was invented by William Britain in 1893. In 1930 a miniature garden range was introduced that included cold frames and flowerpots, a summer house, a pergola and an array of different flowers that could be slotted into flower beds by budding garden designers. Item 043 in the range was the snowdrop, which could be used with crocuses and hyacinths to create a springtime feel. Production of the lead garden items stopped in 1941 and since these old products do not conform to modern toy safety standards, today they are collectors' items only.

Snowdrops feature on many other collectors' items, such as stamps and postcards. Cigarette cards were issued by tobacco manufacturers to stiffen the packaging and promote their brands; in the UK, W. D. & H. O. Wills was one of the first companies to include advertising cards, from 1887. Their set of fifty Old English Garden Flowers cards, issued in 1913, included the snowdrop. The American cigarette brand Old Judge was owned by Goodwin & Co., which produced large numbers of trading cards to promote its brands – initially as sepia-toned photographic albumen prints and later as chromolithographs. Their flower series, printed in Philadelphia in 1890, also includes a snowdrop (*Galanthus nivalis*). Some early cigarette cards were printed on silk which was then attached to a paper backing. Kensitas silk cigarette cards were issued in 1934 and 1935 by the cigarette company J. Wix & Sons of London. They produced an attractive series of flower silks that included beautiful blue delphiniums and intense scarlet anemones. Snowdrops feature on their own on one silk, and on another in a bouquet with crocuses and a primula. Silk cards

Cigarette cards, including Wills's Cigarettes's Old English Garden Flowers series, 1913.

were discontinued in order to preserve resources during the Second World War.

Public interest in cigarette cards led to other producers copying the idea. The series of thirty Devon flowers cards were produced in 1927 by the confectioners James Pascall Ltd. Anyone who managed to collect the complete set could send the cards to the company to receive a presentation casket of sweets and chocolates. James Pascall (1838–1918) worked as an agent for Cadbury's chocolate company before setting up his own confectionery business with his brother Alfred. They had a large factory at Valentine Place in Blackfriars, London, and produced a wide range of different cards to promote their products.

The English tea company Brooke Bond, founded by Arthur Brooke in 1845 (the name reflected the fact that Brooke considered it his

American first day cover of a snowdrop, 1996.

'bond' to provide a quality product to his customers), issued more than 200 series of picture cards to promote their products between 1954 and 1999. Collectors could get an album in which to stick their cards. The company's third wildflowers series, issued in 1964 and which featured snowdrops, was painted by Charles Frederick Tunnicliffe, a member of the Royal Academy best known for his bird paintings. Tunnicliffe (1901–1979) spent most of his life on the Isle of Anglesey off the northwest coast of Wales and much of his work remains on display there at the Anglesey museum Oriel Ynys Môn near Llangefni.

four

Words and Music

> There it stood, so delicate and so easily broken, and yet so strong in its young beauty; it stood there in its white dress with the green stripes, and made a summer.
>
> HANS CHRISTIAN ANDERSEN, 'The Snowdrop' (1863)

In Rebecca Gethin's poem 'Language Garden' (2013), she compares the open petals of snowdrops to musical notes on a stave and the new shoots to punctuation marks.[1] To her, the flowers have words to say: 'in the spring the flowerbeds / fill with sentences'. This ability of plants to convey a cultural dimension, to have something to say to us, is part of the explanation for that deep emotional response that people can experience when contemplating a flower.

Naming one's child after a flower must be one of the most emotional responses a person can have. Snowdrop as a personal name is not as common as many other flower names, such as Rose, Lily and Daisy. In the United States, Rose appeared regularly in the top twenty most popular female names in the first decades of the twentieth century, according to data from the U.S. Social Security Administration.[2] In 2013 Poppy was a new addition to the list of top ten girls' names in the United Kingdom.[3] Although Snowdrop was first used as a personal name back in the nineteenth century, it has always been a rare choice. Birth records for England and Wales from 1837 to 2006 show 185 girls with a given name of Snowdrop, the first being Hannah Sarah Snowdrop Jones of Anglesey, who was born in 1843. In recent years the name is still seen, albeit irregularly. Most babies registered with the name Snowdrop were, not surprisingly, born in the first quarter of the year, when snowdrops are in flower. *The Standard* newspaper in 1892 ran a series of readers' letters discussing the use of flower names, which included the comment from a Philip Bartlett that: 'There is a strong prejudice existing in some minds against naming children after flowers on the ground that children so called are supposed, like the flowers, to be short lived.' This would have been a significant concern in Victorian times, when infant mortality rates were high – and perhaps the snowdrop, with its perceived fragility, was felt to be a particularly risky choice.

The Welsh word for snowdrop, *eirlys*, can also be used as a girl's name. *Kardelen*, the Turkish equivalent, which literally translates as 'snow-poker', is used more frequently as a name for both males and females. In Georgia the word for the snowdrop, *endzela*, is also a girl's name. Unusually, in the Georgian language the word for snow, *tovli*, is very different to the word for the flower. The Italian for snowdrop, *bucaneve*, is not usually used as a personal name but is famous as a pretty, flower-shaped biscuit decorated with white icing. Produced since the early 1950s by the Doria company in Orsago, north of Venice, the biscuits are popular at breakfast dipped into hot milk. In Portugal snowdrops, or *sonjos*, are small meringue cookies made of beaten egg whites, sugar and lemon juice.

The most familiar use of the name Snowdrop in the nineteenth century was that in the famous fairy tale featuring a beautiful girl and seven dwarfs. The best-known version of the story was written by the brothers Jacob and Wilhelm Grimm and was based on the tale told to them by three sisters, Jeannette, Marie and Amalie Hassenpflug of Kassel in central Germany.[4] In the Grimms' original tale, first published in 1812, it was the heroine's own mother who was jealous of her beauty and plotted to kill her, but in a later version in 1819 the brothers edited the story so that her mother died in childbirth

M. Sowerby
SNOWDROP
IN·WINDY, SNOWY, WINTRY HOURS,
OUT OF THE GROUND I PEEP;
QUITE WIDE-AWAKE, WHILE OTHER FLOWERS
AS YET ARE FAST ASLEEP.

Snowdrop postcard, *c.* 1918.

Bucaneve: snowdrop biscuits from Italy.

and a wicked stepmother became the villain. The title of their story, 'Schneewittchen', translates from the German as 'Snow White', but in the first English translation in 1823, Edgar Taylor changed the name to 'Snow-Drop', and it was by this name that the story first became familiar to English-speaking readers. The story was illustrated by artists such as Arthur Rackham (1867–1939), Warwick Goble (1862–1943) and Marianne Stokes (1855–1927). Stokes was born in Austria and moved to Cornwall with her husband in 1886. Her tempera drawing of Snowdrop lying in her glass coffin was exhibited at the Royal Academy Winter Exhibition in 1923 and featured in *The Sphere* newspaper in the same year. The original is now in the Wallraf-Richartz Museum in Cologne. In around 1901 Stokes also painted a charming picture of a young woman reading her prayer book on Candlemas, which is now owned by the Tate in London. Today it is the name Snow White, rather than Snowdrop, that is widely known and used for the fairy tale, due to the pervasive effect of the Walt

Disney animated film *Snow White and the Seven Dwarfs*, which was first released in 1937 and remains much loved.[5]

Snowdrop is a popular name for domestic pets, particularly those with pure-white fur or feathers, including the white kitten washed assiduously by Alice's cat Dinah in Lewis Carroll's novel *Through the Looking-Glass and What Alice Found There* (1871). Snowdrop takes on the role of the White Queen in Alice's dream. Carroll is thought to have chosen the name Snowdrop because it was the name of a kitten belonging to Mary MacDonald, the daughter of his friend and fellow writer George MacDonald.[6]

When in 1850 the American writer Nathaniel Hawthorne moved to a cottage in the Berkshires in Massachusetts, he owned a flock of chickens with floral names including Snowdrop and Crown

Arthur Rackham, illustration of Snowdrop (Snow White) being found by the Seven Dwarfs, 1909.

John Tenniel, 'Snowdrop, My Pet!', from Lewis Carroll's *Through the Looking-Glass* (1871).

Imperial.[7] His novel *The Marble Faun*, written in 1859, includes a chapter titled 'Snowdrops and Maidenly Delights', in which the sculptor Kenyon models 'a beautiful little statue of maidenhood gathering a snowdrop' but feels that it is too fragile a thing to be immortalized in marble.

Beatrix Potter, the author of many much-loved children's books including *The Tale of Peter Rabbit* (1902), had a grey pony 'who was rather lazy' called Snowdrop when she was a child.[8] In a letter replying to children who had written asking about her garden flowers, she reported, 'soon after Christmas we have snowdrops, they grow wild and come up all over the garden and orchard and in some of the woods.'[9] Her gardens at Castle Cottage and Hill Top, in Near Sawrey in the Lake District, provided the inspiration for many of Potter's paintings. An undated watercolour of snowdrops by Potter exists, but they were illustrated in only one of her books, *Peter Rabbit's Almanac for 1929*. In the painting used in the almanac two rabbits are struggling to walk against the wind between lines of snowdrops. Potter ascribed her style of gardening to the presence of the

PUNCH OR THE LONDON CHARIVARI—DECEMBER 19 1945
E. H. Shepard
THE SNOWDROP
"All right, I'll come quietly. But remember I *did* join up in '39"

E. H. Shepard, 'The Snowdrop', cartoon from *Punch*, 1945. 'Snowdrops' was a nickname for members of the U.S. Army's Military Police, who wore white helmets.

snowdrops: 'That is why I have an untidy garden. I won't have the dear things dug up in summer, they are so much prettier growing in natural clumps, instead of being dried off and planted singly.'[10]

Two Royal Navy ships have been given the name HMS *Snowdrop*. The first was launched in 1915 and served during the First World War; the second, a Flower-class corvette, was launched in 1941, served during the Second World War and was scrapped in 1949. There is also a MV *Snowdrop* ferry in operation on the River Mersey in north-west England. Previously known as MV *Woodchurch*, she was refitted and relaunched in 2004 and was renamed as a link to the traditional local flower names, such as Iris and Daffodil, which have been given to the Mersey ferries.[11]

In England it might of course be expected that there are a number of pubs named The Snowdrop. The Snowdrop Inn in Lewes,

Sussex, contrary to expectation, is named not after the flower but to commemorate a tragedy. A huge mound of snow had built up on a chalk cliff just outside the town over the Christmas period in 1836. On 27 December it collapsed, crushing a row of cottages. Eight people lost their lives, and the Lewes avalanche is still the deadliest on record in Britain.[12]

The term Snowdrops was used as a nickname by British civilians for the U.S. Army's Military Police corps, in reference to the white helmets and gloves that formed part of their uniform.[13] A cartoon titled 'The Snowdrop', printed in *Punch* in December 1945, shows a military policeman arresting an injured soldier in Bretton Woods with the caption 'All right, I'll come quietly. But remember I did join up in '39.' This is a reference to the Bretton Woods Conference, held not actually in woods but in the plush Mount Washington Hotel in rural New Hampshire, at which 730 delegates from the 44 Allied nations met to regulate international monetary policy after the Second World War.[14] The *Punch* cartoon was drawn by the renowned artist and book illustrator Ernest Howard Shepard (1879–1976), who is known for his illustrations for the children's books *Winnie-the-Pooh* by A. A. Milne and *The Wind in the Willows* by Kenneth Grahame.

The name Operation Snowdrop has been given to several military winter manoeuvres and exercises. In 1948 U.S. Army paratroops at Pine Camp in New York worked in arctic conditions in Operation Snowdrop, delivering supplies and helping stranded civilians caught in the freezing weather. There was also a military operation by this name that delivered food and medical supplies to snowbound communities in Scotland during the hard winter of 1954–5. Nearly three hundred flights of RAF aircraft and Royal Navy helicopters dropped animal fodder and other supplies to farms and villages, from Inverness-shire to Shetland.[15]

The word 'snowdrops' has been used in Russia to describe human corpses revealed by the melting snow. This idea is used in the thriller *Snowdrops* (2011) by A. D. Miller, which is set in modern

Russia. Miller worked as a foreign correspondent in Russia and his novel looks at corruption in the Russian and expat communities in Moscow, using snowdrop corpses as a metaphor for the way that past experiences cannot always be repressed. The novel was shortlisted for the Man Booker Prize in 2011.

Some people describe the activity of visiting gardens in late winter to view the snowdrops as 'snowdropping'. This word has an alternative meaning, however, having been used to refer to the theft of underwear from washing lines. This was described by an Australian reader, 'Delta', in a letter to the editor of the *Sydney Morning Herald* of Wednesday 12 March 1930:

> Fifty or sixty years ago when public laundries were few, people did their own washing, and it was the practice for two or three young men with more energy than discretion to venture out on a clear moonlight night and go 'snow-dropping', that is, they would rob some clothes line of the weekly washing and steal the snow-white linen that was spread out on the grass to bleach. In those far-away days careful householders left the washing on the line and spread the garments on the grass as the clear atmosphere at night would bleach them clean and white.

Variations of the so-called 'language of flowers', in which meanings are ascribed to particular species of plants, have been practised for thousands of years. Examples occur in the Bible and are frequent in the works of Shakespeare, with rosemary for remembrance, as spoken by Ophelia in the play *Hamlet* (1603), perhaps being the most widely known: 'There's rosemary, that's for remembrance. Pray you, love, remember. And there is pansies, that's for thoughts' (Act IV, Scene 5). Interest in the language of flowers became especially widespread during the late eighteenth and early nineteenth centuries, stimulated in part by misinterpretations of the Turkish *sélam*, a poetical art of using flowers in expressing emotions.[16] The letters from Turkey written by Lady Mary Wortley Montagu (1689–1762), wife of the British Ambassador, reported that a coded language was used by women in the harems. Montagu was a close observer of the life of women in the Ottoman Empire and was noted for having her children inoculated with the smallpox virus after she witnessed the immunity of Turkish milkmaids who had been treated in this way. In a letter to a friend dated 1718, Lady Montagu enclosed a 'Turkish love letter' and included a list of items such as pearls and flowers, citing meanings for them. The examples she gave included jonquil, which she wrote means 'have pity on my passion', and rose, meaning 'may you be pleased, and all your sorrows be mine'. Montagu added:

> There is no colour, no flower, no weed, no fruit, herb, pebble or feather that has not a verse belonging to it; and you may quarrel, reproach or send letters of passion, friendship or civility, or even of news, without ever inking your fingers.[17]

The *sélam* was actually a system of memorization in which the names of flowers rhyme with standard lines of poetry and act as an aid by which the lines could be learned. In the 1839 edition of *The Language of Flowers; with Illustrative Poetry* (1934) the editor Frederic Shoberl gave the example of the pear: 'the word armonde (pear) rhymes among other words with omonde (hope) and this rhyme is filled up as follows: Armonde: Wer banna bir omonde: Pear, Let me not despair.' Mention was also made of the system in Aubry de La Mottraye's account of 1723 of his visit to the court of Charles XII of Sweden at Bender in Moldova, then under Turkish control, at the time of the king's exile after the Battle of Poltava in 1709.[18] La Mottraye spent 26 years travelling from Scandinavia to the Middle East and Africa. The book that resulted from his travels was very successful, combining good writing with quality illustrations by artists such as the English painter and satirist William Hogarth (1697–1764), whose thirteen folio prints for Mottraye's work were some of his first commissions. The text gives examples of verses which 'the young Girls learn by Tradition of one another'. The *Abécédaire de flore, ou langage des fleurs* (Alphabet of Flowers; or, Language of Flowers, 1810), by a French writer named B. Delachénaye and dedicated to Empress Marie Louise of France, used this idea. The author created a phonetic language based on the pronunciation of flower names, rather than simply ascribing esoteric meanings to plants. Birds and insects were used as punctuation.

Many floral dictionaries were published that disseminated the idea that each flower has a particular meaning. Some of the first were seen in France; as well as Delachénaye's, they included Alexis Lucot's *Emblemes de flore et des végétaux* (Emblems of Flora and Plants) and Charlotte Latour's *Le Langue des fleurs* (The Language of Flowers), both published in 1819. In England, Saunders & Otley published *Floral Emblems* by Henry Phillips in 1825 as well as Shoberl's *The Language of Flowers* in 1834. Shoberl's book was in fact a translation of Latour's work, with additions and alterations. Catharine Waterman published *Flora's Lexicon* in Boston in 1839, and in the 1855 edition wrote in the preface: 'The Language of Flowers has recently attracted so much attention, that an acquaintance with it seems to be deemed, if not an essential part of a polite education, at least a graceful and elegant accomplishment.' Many popular magazines and newspapers referenced the language of flowers and the phrase 'say it with flowers' would have had much more meaning for readers at the time than it does today.

Copyright laws were still being developed at this time. In Britain, the Statute of Anne of 1710, named after Queen Anne, was the first copyright statute and granted publishers of a book legal protection for fourteen years. The Engravers' Copyright Act, passed in 1735, was prompted by William Hogarth, whose work was heavily plagiarized.[19] The fact that many of the same lists of flower definitions appear in a large number of similar books from the nineteenth century indicates that much copying of material still occurred.

Snowdrops feature regularly in these flower dictionaries. Shoberl wrote of the snowdrop, 'This firstling of the year may not inaptly be considered as an emblem of hope. Some have regarded it as a symbol of humility, of gratitude, and of virgin innocence.'[20] Waterman's *Flora's Lexicon* included the snowdrop as a message of consolation. *Flowers and their Kindred Thoughts*, published in London in 1848, was designed as a gift book by Owen Jones, a Welsh architect and influential designer. He produced some of the earliest chromolithographic books, several of which were illustrated by Edward La Trobe Bateman. *Flowers and their Kindred Thoughts* brought together fourteen chromolithographs of flowers with poems by Mary Ann Bacon printed in gilt lettering. The double snowdrop *G. nivalis* 'Flore Pleno' was illustrated to represent hope. Bacon's verse includes the lines,

> I've seen the snow lie deep upon the earth
> Blotting my garden home. I've felt how grief
> Can check the sources of the spirit's mirth
> Stopping all nature's inlets to relief.

In the book *Language of Flowers* (1884) by the popular English author and illustrator Kate Greenaway the snowdrop is again recorded as an emblem of hope. Its habit of flowering at the tail-end of winter in what is often inclement weather encourages the belief that things will get better.

Snowdrop flowers often appear in poetry and prose. In the Hans Christian Andersen fable 'The Snowdrop' of 1863 a flower stirs beneath snowy ground and strives to reach the summer. It is greeted by sunbeams and is praised for being the first flower, but is then mocked as a 'summer gauk' or fool by the wind and weather. Picked by a little girl, the flower is pressed and sent with joke verses to a friend in the manner of a Danish valentine. A maid accidentally transfers the snowdrop to the pages of a book of poetry by Ambrosius Stub, whom Andersen considered, like the snowdrop, to be a creature before his time. The verses of the Danish poet Stub (1705–1758) had a strong lyrical character and often conveyed a religious message. Though only six of his works were published in his lifetime Andersen was not his only admirer, and Stub's collected works were published posthumously in 1771.

The physician and political writer John Shebbeare was best known for his *Letters on the English Nation* (1756). In his book *Lydia; or, Filial Piety: A Novel* (1755) he used the snowdrop as a way of commenting on the character of his heroine: 'She now dressed herself as clean as the Snow-drop or the variegated Gold-finch.' The goldfinch *Carduelis carduelis* occurs often in paintings, probably most famously in the small painting of 1654 by Carel Fabritius that in 2013 was made the subject of the successful novel by Donna Tartt. The ornithologist Herbert Friedman found 486 examples of goldfinches included in Renaissance paintings, attributed to 254 different artists;[21] it is usually shown being clutched by the infant Jesus and is used as a symbol of the soul, of resurrection and of humans' potential for redemption. By comparing Lydia to a snowdrop and a goldfinch, Shebbeare is emphasizing the purity of her soul.

In George Eliot's first novel, *Adam Bede* (1859), Eliot describes how 'Lisbeth Bede loves her son with the love of a woman to whom her first-born has come late in life. She is an anxious, spare, yet vigorous old woman, clean as a snowdrop.' Eliot felt that realism was very important in novels and endeavoured to use physical details and local voices to help bring to life her characters' internal thoughts and motivations.[22] Queen Victoria was an avid reader of Eliot's novels and enjoyed *Adam Bede* so much that she commissioned the artist Edward Henry Corbould to paint scenes from the book.[23]

Snowdrops feature in Charles Dickens's novel *Barnaby Rudge* (1841), in which Mr Willet scolds Joe for picking winter flowers:

> 'And what do you mean by pulling up the crocuses and snowdrops, eh sir?'
>
> 'It's only a little nosegay,' said Joe, reddening. 'There's no harm in that, I hope?'

Shirley (1849), a novel set in the years 1811–12 against a backdrop of the Luddite uprisings in Yorkshire, was Charlotte Brontë's second published novel after *Jane Eyre*. In it the hero Robert Moore gathers a winter bouquet for Caroline Helstone:

> A sweet fringe of young verdure and opening flowers – snowdrop, crocus, even primrose – bloomed in the sunshine under the hot wall of the factory. Moore plucked here and there a blossom and leaf, till he had collected a little bouquet; he returned to the parlour, pilfered a thread of silk from his sister's work-basket, tied the flowers, and laid them on Caroline's desk.

In *Jane Eyre* Brontë describes how Jane delights in the arrival of snowdrops among other spring flowers when she is at school at Lowick. Brontë herself, on the occasion of her marriage on 19 June 1854, was described by Sutcliffe Sowden as 'a snowdrop, a pale wintry flower'. Sowden, who was the vicar of St James's church in Hebden Bridge and a friend of the groom, Arthur Bell Nicholls, had travelled to Haworth to perform the marriage ceremony. Brontë became pregnant soon after the wedding but sadly she and her unborn child died in March 1855.

The American novelist Louisa May Alcott (1832–1888) was born in Germantown, Pennsylvania. Alcott's family moved to Boston in 1838, where her father started a school and joined the Transcendental Club with Ralph Waldo Emerson and Henry David Thoreau. Founded in 1836, the Transcendental Club was a forum for new ideas and encouraged the belief that organized religion and political parties corrupted the purity of the individual. Alcott and her sisters had to work from an early age to help support the family. Her first book was *Flower Fables* (1849), a selection of tales that she originally wrote for Ellen Emerson, the daughter of Ralph Waldo Emerson. *Little Women*, published to great acclaim in 1868, was a semi-autobiographical novel based on Alcott's childhood.[24]

When the character Beth, modelled on Alcott's sister Lizzie, dies, snowdrops flower as a consolation:

> But a bird sang blithely on a budding bough, close by, the snowdrops blossomed freshly at the window and the spring sunshine streamed in like a benediction over the placid face upon the pillow, a face so full of painless peace that those who loved it best smiled through their tears and thanked God that Beth was well at last.

In a sequel to *Little Women*, *Jo's Boys* (1886), snowdrops now symbolize fresh hope when news comes that Emil, who had been thought lost at sea, was safe: 'best of all, little Josie lifted up her head as the snowdrops did, and began to bloom again, growing tall and quiet.'

Written at a similar date to Alcott's work, the stories of Sheridan Le Fanu (1814–1873) are of a very different style. Born in Dublin, Le Fanu wrote psychological ghost stories and Gothic novels featuring female vampires, including the collection *In a Glass Darkly* (1872). In *The House by the Churchyard* (1863) snowdrops are laid on the dead body of Lily, like those that bloom for Beth in Alcott's novel:

> and there was little Lily, never so like the lily before. Poor old Sally had laid early spring flowers on the white coverlet. A snow-drop lay by her pale little finger and thumb, just like a flower that has fallen from a child's hand in its sleep.

In the poem 'A Backward Spring', written during the First World War in April 1917, Thomas Hardy uses personification, describing trees fearful of opening their buds. The snowdrop, however, is not discouraged by the lengthy winter, which may be interpreted as a sign of hope or, conversely, an indication of nature's indifference to man:

Galanthus 'Robin Hood' snowdrop showing pale green 'hieroglyphs' on petals.

Anna Laetitia Barbauld (second from left, standing), in Richard Samuels's *Portraits in the Characters of the Muses in the Temple of Apollo*, 1778, oil on canvas.

'Yet the snowdrop's face betrays no gloom, / And the primrose pants in its heedless push'. Hardy's fourth novel, *Far from the Madding Crowd*, originally appeared in 1874 as a monthly serial and was immediately successful. In it he demonstrates the romanticism of his character Troy, who, remorseful after the death of his true love, Fanny, spends all his money on a marble tombstone for her and plants snowdrops and other flowers on her grave:

> He hung his lantern on the lowest bough of the yew-tree, and took from his basket flower-roots of several varieties. There were bundles of snowdrop, hyacinth and crocus bulbs, violets and double daisies, which were to bloom in early spring, and of carnations, pinks, picotees, lilies of the valley, forget-me-not, summer's farewell, meadow-saffron and others, for the later seasons of the year.

> Troy laid these out upon the grass, and with an impassive face set to work to plant them. The snowdrops were arranged in a line on the outside of the coping, the remainder within the enclosure of the grave. The crocuses and hyacinths were to grow in rows; some of the summer flowers he placed over her head and feet, the lilies and forget-me-nots over her heart.

Walter de la Mare won the James Tait Black Memorial Prize for fiction for his fictional account *Memoirs of a Midget*, written in 1921. Among the details of the unsettling world of Miss M, the small heroine of the tale, is the description of a conversation that could take place between any modern snowdrop lovers: she describes how Mr Crimble 'stood nervously twitching a small bunch of snowdrops which he assured me were the first of the New Year. I thanked him and remarked that our Lyndsey snowdrops were shorter in the stem than these and had he noticed the pale green hieroglyphs on the petals?'

This extract from the poem 'The Invitation' (1825) by the poet and writer Anna Laetitia Barbauld, despite denying the lovely honey scent that many snowdrops have, was very highly regarded in its time and is much quoted in contemporary collections of poems for children and botanical books:

> Already now the snow-drop dares appear,
> The first pale blossom of the unripened year;
> As Flora's breath, by some transforming power,
> Had changed an icicle into a flower;
> Its name and hue the scentless plant retains,
> And Winter lingers in its icy veins.

Barbauld was born in Leicestershire, where she had little female companionship and persuaded her father to teach her Latin and Greek, which at the time were really not considered suitable subjects for young ladies.[25] The family moved to Warrington, where Barbauld became friends with Joseph Priestley (1733–1804), the theologian and the first person to isolate the gas oxygen. Inspired by Priestley's attempts at writing verse, Barbauld started writing her own poetry.[26] She married Rochemont Barbauld in 1774 and together they established a boarding school in Palgrave, Suffolk; she continued to write prolifically, including hymns for children and a number of essays about political and social concerns. Her poetry was admired by other distinguished writers, including Oliver Goldsmith, Samuel Taylor Coleridge and William Wordsworth – although in later life Wordsworth wrote that she 'was spoiled as a poetess by being a dissenter, and concerned with a dissenting academy.'[27]

The writer Marguerite Gardiner, Countess of Blessington, was born in Ireland and published her first book, *The Magic Lantern*, in 1822. A portrait of her from that year by Thomas Lawrence was exhibited at the Royal Academy and 'set all London raving', according to Byron. It has been described as 'an exquisite painting, at once voluptuous and touching – the Countess wearing her beauty as though it is as much a burden as an asset'.[28] At age fourteen she had been forced to marry an army captain, who was physically abusive to her. In 1823 Blessington travelled for several weeks in Italy with her third husband, John Blessington, and with Lord Byron, leading to her most famous work, *Conversations with Lord Byron*, published ten years later. Blessington was commissioned by the noted publisher and lithographer Rudolph Ackermann to write the poetry for the book *Flowers of Loveliness*, which was to be a gift book for women, to capitalize on the current craze for flower and poetry books. The plates were by the artist Edmund Thomas Parris, who produced twelve groups of female figures designed to be emblematic of flowers. Blessington's accompanying poems examined the idea that women can express the essential qualities inherent to flowers: 'In Flowers and Blossoms, Love is wont to trace, / Emblems of Woman's virtues and her grace'. She pointed out that she herself was named for a flower, 'The English daisy is called Marguerite in France'. Parris's design for the snowdrop

Thomas Lawrence, *Countess of Blessington*, 1822, oil on canvas.

flower shows three sisters at the grave of their mother, with the youngest clutching a handful of snowdrops. Blessington's poem is about a child coming to terms with her mother's death and considers the innocence of a young child and ability of women to support each other through grief:

> Here's a flower – the first I found,
> 'Gainst she wakes – she loves it well.

Edmund Thomas Parris, 'Snowdrop', illustration from *Flowers of Loveliness: Twelve Groups of Female Figures Emblematic of Flowers* (1866).

Ah! How still is all around!
Will she waken? – Sister, tell.

The epic poem *Os Lusíadas* (The Lusiads) by Luís Vaz Camões is regarded as Portugal's national poem; it was published in 1572 after the author returned from travelling in the Far East. It takes as its theme the Portuguese maritime adventures of the fifteenth and sixteenth centuries, and in it the poet described damsels' cheeks as 'paler than the hue of snowdrops trembling to the chilly gale'.

Many of the great Scottish writer Robert Burns's poems celebrate pastoral pleasures. In 'My Nanie's Awa', a song written in 1794, he celebrates the first flowers of spring: 'The snawdrap and primrose our woodlands adorn, / And violetes bathe in the weet o' the morn'. In 'The Parting Kiss' (1788), Burns compares a snowdrop to a kiss:

> Humid seal of soft affections,
> Tend'rest pledge of future bliss,
> Dearest tie of young connections,
> Love's first snow-drop, virgin kiss.

The English writer Mary Robinson (*c*. 1757–1800) began work as a teacher at the age of fourteen at the school that her mother ran; later, a chance meeting with the actor David Garrick led to an interest in acting. When her father closed the school she was married off to Thomas Robinson, an alcoholic gambler who was sent to debtors' prison in 1775. Mary and her daughter lived with him at King's Bench Prison for several months. At the age of 21 she became the mistress of the Prince of Wales (the future George IV), but he lost interest in her after a year and she wrote poetry out of financial necessity. Robinson's poem 'The Snowdrop' was originally published in her novel *Walsingham; or, The Pupil of Nature* in 1797. It describes the snowdrop as a thing of beauty battered by the harshness of winter, reflecting Robinson's own life:

> The Snow-drop, Winter's timid child,
> Awakes to life bedew'd with tears;
> And flings around its fragrance mild,
> And where no rival flow'rets bloom,
> Amidst the bare and chilling gloom,
> A beauteous gem appears!

Robinson knew many other authors, including Wordsworth and the feminist writer Mary Wollstonecraft, and was a personal friend of Coleridge, a fellow contributor to the *Morning Post* newspaper. The manuscript for Coleridge's 'The Snow-drop' of 1800 is headed:

'Lines written immediately after the perusal of Mrs Robinson's "Snow Drop"'. While ostensibly about the flower, Coleridge's poem uses the snowdrop as a metaphor for Robinson herself:

> Fear no more, thou timid Flower!
> Fear thou no more the winter's might,
> The whelming thaw, the ponderous shower,
> The silence of the freezing night!

Wordsworth himself, though best known for his love of daffodils, wrote in praise of the snowdrop in his poem 'On Seeing a Tuft of Snowdrops in a Storm' (1820) of

> . . . these frail snow-drops that together cling,
> And nod their helmets, smitten by the wing
> Of many a furious whirl-blast sweeping by.

His second snowdrop poem, 'To a Snowdrop', written around the same time, has stimulated speculation that it may be an oblique reference to Robinson.[29]

> Lone Flower, hemmed in with snows and
> white as they,
> But hardier far, once more I see thee bend
> Thy forehead, as if fearful to offend,
> Like an unbidden guest . . .

There are parallels between the lives of Mary Robinson, Marguerite Gardiner and the Romantic poet Charlotte Turner Smith. Smith (1749–1806) was forced into an early marriage to an abusive husband and spent time with him in debtors' prison. She wrote her first book of poetry while in King's Bench Prison, the payments for which facilitated the release of her husband. She later left him and used her writings to support their twelve children. Smith is

Thomas Gainsborough, *Mary Robinson*, 1781, oil on canvas.

said to have initiated the revival of the English sonnet and was the author of ten political novels and works of Gothic fiction. Her poetry was praised by Wordsworth, Coleridge and Walter Scott. The poem 'Snowdrops' from her collection *Elegiac Sonnets and Other Poems, Volume II* (1797) links the flowers with grief:

> Ah! ye soft, transient children of the ground,
> More fair was she on whose untimely grave
> Flow my unceasing tears! . . .

The English poet John Clare (1793–1864) was much inspired by the countryside that surrounded him in the Northamptonshire village of Helpston. His poem 'The shrill bat there its evening circles makes' describes many familiar, early flowering plants:

> The gold-eyd daisey with its ruddy stain
> Will even venture ere the frosts are bye
> & on the snow drops tiny couch remain
> & neath a privet hedge soft sheltering nigh
> The violet often blooms, nor waits an April sky.

'The Sensitive Plant' was written by the renowned lyrical English poet Percy Bysshe Shelley following the death of his child while his wife, Mary Shelley, was suffering from depression. It was published in *Prometheus Unbound and Other Poems* in 1820 and includes the verse:

> The snowdrop, and then the violet,
> Arose from the ground with warm rain wet,
> And their breath was mixed with fresh odour, sent
> From the turf, like the voice and the instrument.

The poem marks the progression of the seasons from spring to winter and was written as an allegory of the difficulty of determining one's place in a universe that is changing. The sensitive plant of the title is *Mimosa pudica*, whose leaves respond rapidly to touch, and this was used by Shelley as a metaphor for himself.[30]

Elizabeth Barrett Browning may not have been a fan of snowdrops but Robert Browning, in his wistful poem 'The Lost Mistress' of 1845, imagines a lover accepting the end of an affair but with a huge sense of loss: 'Your voice, when you wish the snow-drops back, / Though it stay in my soul for ever!' A less soulful reference occurs in *Pacchiarotto* (1876), in which Browning pleads, 'Don't trample the grass – hocus-pocus / With grime my Spring snowdrop and crocus.'

The essayist and poet Matthew Arnold was also familiar with snowdrops, and in one of his most famous poems, *Tristram and Iseult* of 1852, he compares the princess Iseult to the flower:

> Who is this snowdrop by the sea? –
> I know her by her mildness rare,
> Her snow-white hands, her golden hair;
> I know her by her rich silk dress,
> And her fragile loveliness.
> The sweetest Christian soul alive,
> Iseult of Brittany.

On 2 March 1871 one of the Victorian era's most famous poets, Gerard Manley Hopkins, wrote to his mother to wish her a happy fiftieth birthday from Lancashire, where it was as 'dank as ditchwater . . . However it is not cold, birds are singing, and the garden is full of clumps of snowdrops.'[31]

Oscar Wilde's poem 'Ravenna', which reflects on a visit to Italy in March 1877, was dedicated to his friend George Fleming. In it Wilde remembers 'all the flowers of our English Spring, / Fond snowdrops, and the bright-starred daffodil.'

The English poet Olive Custance became part of the Aesthetic movement in the 1890s; in 1902 she eloped to marry Lord Alfred

Douglas, of whom her father did not approve, as Douglas had been the friend and lover of Oscar Wilde.[32] Custance published in the quarterly magazine *The Yellow Book*, an important journal of the time – Aubrey Beardsley was its first art editor and contributing authors included Max Beerbohm and William Butler Yeats. Custance's poem 'The Waking of Spring' featured in 1895 and includes a reference to snowdrops: 'And in the woods stand snowdrops, half asleep, / With drooping heads sweet dreamers so long lost.'

The renowned poet and children's writer Ted Hughes was Poet Laureate from 1984 until his death in 1998. His great interest in the natural world suffuses much of his poetic work. Hughes's second collection of poems, *Lupercal* (1960), contains the powerful poem 'Snowdrop', in which winter is viewed from the perspective of a hibernating mouse. The feminine snowdrop is identified only by the poem title and Hughes emphasizes the conflict between the fragility of her pale head and her ability, 'Brutal as the stars', to defeat winter. Hughes's children's science fiction book *The Iron Woman* (1993), which was a sequel to his more famous *The Iron Man* (1968), is a cry against the dumping of industrial waste in the rivers and seas of the world. The snowdrops that the Iron Woman lets fall to earth are used again to symbolize cleanliness and purity.

Snowdrops continue to inspire poetry, and although the forms of the poems and the words may change, the topics of death and remembrance are a constant. Margaret Wilmot was born in California but has lived in Sussex since 1978. Her poem 'My Mother's Sleep is Deep' was published in 2013. It compares her dead mother's bones to the whiteness of winter flowers.

The *manhwa* (Korean comic) book *Snow Drop* by Kyung-ah Choi was first published in Korea in 1992, with an English translation by Jennifer Hahm published in 2004. It tells the story of So-na, whose name means pine tree, and the nursery called Snowdrop that was her mother's love. So-na tells a customer about the story of the snowdrop being created as a consolation for Eve.

There has been speculation that the snowdrop was the magical herb 'moly' that grew from the blood of Gigante in Book Ten of Homer's *Odyssey*.[33] Hermes gave the herb to Odysseus to provide immunity to the magic of the enchantress Circe. The herb was described as having a black root, while the flower was as white as milk. It was meant to be 'dangerous for a mortal man to pluck from the soil, but not for the deathless gods'. This makes the identification with the snowdrop somewhat unlikely, since snowdrops do not have black roots, although small black fungal masses do sometimes develop on bulbs infected with grey mould. It is also generally not now considered to be dangerous to pluck snowdrops from the soil. If you try to do so without a trowel you are likely to end up with just a handful of flowers and leaves, with the bulb remaining in the ground, which would be frustrating rather than dangerous (unless they happened to be particularly treasured snowdrops belonging to somebody else . . .).

Whether or not the snowdrop is the herb moly, it does have an important role to play in modern medicine. The snowdrop of words and music may be of help to those for whom words are losing their significance. Alzheimer's disease is a horrible condition and is the most common cause of dementia. It was first described by the German neurologist Alois Alzheimer in 1906 and is a progressive disease that affects the neurons in the brain.[34] The first symptoms of Alzheimer's disease can include lapses in memory and a struggle to find the right words, although as most of us experience these symptoms from time to time diagnosis may be difficult and is often delayed. Snowdrops have not been widely used medicinally in Europe and are not mentioned in many of the oldest medico-botanical texts. However, there are limited observational studies in the area of the Caucasus Mountains in Russia and Georgia which report peasant women using a decoction of the bulbs of *Galanthus woronowii* for the treatment of polio in children.[35] A Bulgarian pharmacologist noticed snowdrops being rubbed on the forehead as a folk medicine for headaches. Snowdrops contain a natural alkaloid called galantamine, which was first extracted from *G. woronowii* in 1952, although it is also found in other *Galanthus* species and in *Narcissus* and *Leucojum*.[36]

Galanthus woronowii is a source of the drug galantamine, which is used in the treatment of Alzheimer's disease.

Galantamine has the ability to penetrate the blood–brain barrier and reduce the breakdown of acetylcholine, a natural brain chemical; this can improve cognitive ability in people with Alzheimer's disease. It cannot cure dementia, but it does slow down its progression and may ameliorate the symptoms of memory loss and confusion. Those using these treatments can experience improvements in motivation, anxiety levels and confidence.

Galantamine was first licensed in 2000 for the treatment of Alzheimer's disease in Iceland, Ireland, Sweden and the UK. Guidance from the National Institute for Health and Care Excellence (NICE) in the UK was updated in March 2011 to allow the use of galantamine for people with both mild as well as moderate Alzheimer's disease.[37] Clinical trials have demonstrated the positive effect of the drug, which is sold under a number of trade names, including Acumor, Elmino, Galantex and Zeebral. Commercial production of galantamine has been mainly from wild-collected *Leucojum aestivum* plants; because the isolation of galantamine from plant material is expensive and prone to supply problems, researchers are looking at methods to produce galantamine synthetically.[38]

Music

Snowdrops have inspired music as well as poetry, perhaps most famously the piece *Schneeglöckchen* (Snowdrops Waltz) by Johann Strauss II (1825–1899). The Austrian composer Strauss was born in St Ulrich near Vienna and during his career became known as the Waltz King. This was despite the fact that his father, the conductor and composer Johann Strauss I, wanted him to become a banker, and whipped his son when he caught him studying the violin in secret.[39] Strauss the Younger composed over 400 pieces of dance music and is best known for 'The Blue Danube' and the operetta *Die Fledermaus*. Snowdrops grow wild in the woods that partly surround Vienna.[40] Despite the success of the waltz *Geschichten aus dem Wienerwald* (Tales from the Vienna Woods, 1868), the composer was not thought to be keen on walking in the woods himself. Indeed, he expressed a fear of climbing even gentle hills, which probably caused problems between him and his friend the composer Johannes Brahms, who enjoyed dragging his friends on mountain-climbing expeditions.[41] *Schneeglöckchen*, like the flowers that inspired it, is a piece with great delicacy. First performed in 1834, it has a cello introduction and is sometimes performed with bells to emphasize the wintry feel of the work.

William Smallwood's *The Snowdrop Waltz* sheet music, 1896.

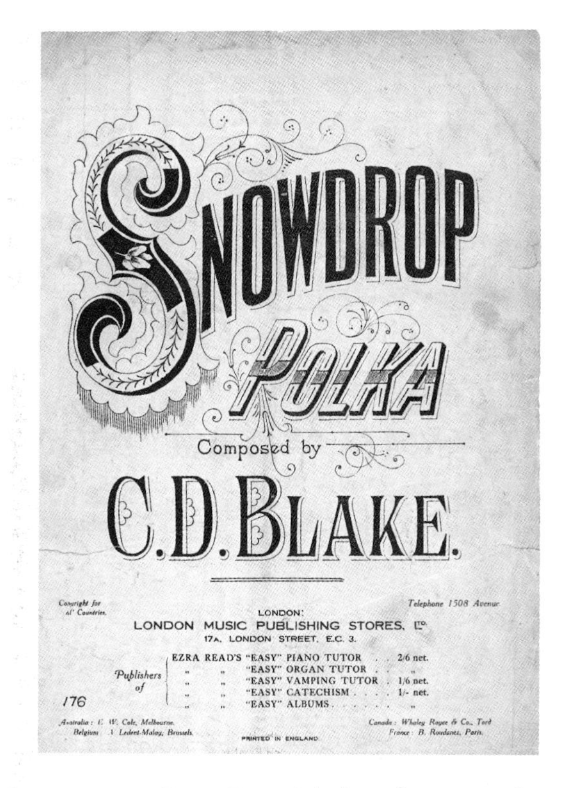

Charles Dupree Blake's *Snowdrop Polka*, 1890.

Another Austrian composer, Franz Peter Schubert (1797–1828), was astonishingly prolific. By his death at the age of 31 he had written over 1,599 works, the majority of which were songs for solo voice and piano. In *Viola* he set words by the poet and actor Franz von Schober (1796–1882) to music; the piece begs a snowdrop to comfort a jilted bride. The German composer Robert Schumann set a poem by Friedrich Rückert (1788–1866) to music in 1849 to create *Schneeglöckchen*, a song for voice and piano that is a tender evocation of spring. Schumann (1810–1856) had intended to build a career as a pianist but an injury to his hand led him to become a composer instead. In the early years of his marriage to Clara Wieck he worked on a large number of songs, including the setting to music of poems by Goethe, Byron and Burns as well as others by Rückert.

Frédéric François Chopin (1810–1849) was one of the most important piano composers of the nineteenth century. Between the years

1835 and 1839 he wrote a set of 24 Preludes, Op. 28, of which the third, 'The Snowdrop', is one of the shortest. Chopin himself was described as 'always ill, and as frail as a snow-drop, but an exquisite genius'.[42] Around the same time in the United States, Estelle Hewitt composed a *Snowdrop Waltz* (published in Baltimore in 1847). James Bellack (*c.* 1814–1891), a prolific composer of sentimental songs and works for piano, also produced a waltz. His *Snowdrop* in C major was published in Philadelphia in 1855. An English *Snowdrop Waltz* was composed in 1896 by William Smallwood, a man of 'genial and generous disposition' from Kendal in Cumbria. He came from a musical family and by the time he was seven he was a competent player of the flute. He was described playing the chapel organ as 'a chubby-faced lad, attired in a round-about jacket and wearing a cap upon his curly head'.[43] He became organist at St George's church in Kendal at the age of fifteen and held the post for fifty years. The food writer Josceline Dimbleby describes in her family memoir *A Profound Secret* (2004) how her great-grandmother May Gaskell danced the *Snowdrop Waltz*, a gallop called 'Fizz' and several other dances with her husband-to-be, Captain Gaskell.[44]

Snowdrop flowers on their long pedicels are often said to dance in the breeze, and they do seem to have inspired a number of pieces of dance music. Nicholas Vickery composed *The Snowdrop* mazurka – a lively Polish folk dance – for the piano. A *Snowdrop Polka Mazurka* was also written by the London music publisher and composer Charles W. Jeffreys (1807–1865), while the *Snowdrop Polka* was written by Charles Dupree Blake (1847–1903) and *The Snow Drop* polka was written in 1859 by George Wilson. Another *Snowdrop Polka* was written by James J. Freeman, who managed to find dance music in far more unlikely subjects, as he also composed a *Telephone Polka*. A *schottische* is a slower form of polka that is thought to have originated in Bohemia, and the *Snowdrop Schottische* was written by Amelia Edwards and published in New York in 1852. Edwards had also composed a *Snowdrop Polka* the previous year, which was arranged by William Dressler.

Tchaikovsky's *The Seasons* is a collection of twelve short piano pieces, each one representing a month of the year. They were commissioned in 1875 by Nikolay Bernard, the editor of the St Petersburg music magazine *Nouvelliste*. Bernard suggested a subtitle for each month's piece, such as 'At the Fireside' for January, 'Song of the Reaper' for July and October's 'Autumn Song'. Each also had an epigraph taken from Russian literature. 'Snowdrop' was chosen as the subtitle for April and the epigraph was by Apollon Maykov (1821–1897), a Russian poet best known for his lyric verse:

> The blue, pure snowdrop – flower,
> and near it the last snowdrops.
> The last tears over past griefs,
> and first dreams of another happiness.

Handful of Snowdrops is the name of an alternative techno rock band from Quebec City that first emerged in 1984. The band split in 1993 but returned to the music scene with its own label, NanoGénésie, in 2009. The band's founder member Jean-Pierre Mercier explained that the name of the band came from an image, carved into a wooden desk used in one of his college classes, of a fist holding snowdrops. He appreciated the resilience of snowdrops in being able to emerge through the ice, describing them as 'small but headstrong'.[45]

five
Collectors and Conservation

> To see a World in a Grain of Sand
> And a Heaven in a Wild Flower
> WILLIAM BLAKE, 'Auguries of Innocence' (1803)

There is an old Italian saying that roughly translates as 'the tale is not beautiful until you add to it.' For many people the pleasure received from a flower is increased when they feel an emotional connection with it. This may be stimulated by something as simple as remembering a walk in a garden or through a snowdrop wood or by standing in front of a painting in an art gallery. Greater involvement may come if one is given a snowdrop to grow, when memories of the donor can increase the emotions surrounding the plant. Many of us first encounter the common snowdrop as a child and it is only later that we realize the number of different species and cultivars that are available. A mild interest in their variability can quickly lead to a growing collection in the garden and from there it is no long step to an obsession with the plant. Sharing plants with friends, swapping plants with fellow addicts or even selecting some new variation and naming it can enable you to become part of another person's snowdrop story.

In George MacDonald's lengthy verse 'A Manchester Poem' (1893) a couple goes walking on their day off from work and find the remains of an old cottage with a single snowdrop flowering in the garden.

> With careful hands uprooting it, they bore
> The little plant a willing captive home;
> Fearless of dark abode, because secure
> In its own tale of light . . .

Today, removing even a single snowdrop may be disapproved of or have legal implications. In Victorian times, however, there was less appreciation of conservation issues. MacDonald was a Scottish clergyman, poet and fantasy novelist whose work influenced many other writers, including C. S. Lewis and J.R.R. Tolkien. The poet draws religious significance from his simple tale and tries to show that the lowly flower 'of harassed spring' has more to say than a stately lily, and 'therefore high hope is more than deepest joy'.

C. S. Lewis, the author of *The Chronicles of Narnia* (1950–56), himself grew snowdrops and wrote a charming letter to some young

A. C. van Eeden, '*Galanthus nivalis*', from *Album van Eeden: Haarlem's Flora* (1872).

fans in January 1954 in which he reported: 'There is no snow here yet and it is so warm that the foolish snowdrops and celandines . . . are coming up as if it was spring.'[1] Snowdrops are more frequently described as brave than foolish.[2] To many people these delicate-looking plants that can survive inclement conditions are the source of great fascination. The exhilaration felt when the plant lover studies an interesting plant may well be dismissed as the ravings of an obsessive by those who do not share the feeling. In *The Story of David Gray*, published in 1900, Robert Buchanan wrote:

> Could you understand
> One who was wild as if he found a mine
> Of golden guineas, when he noticed first
> The soft green streaks in a snowdrop's inner leaves?

Plant obsession is certainly nothing new. The female pharaoh Hatshepsut (1508–1458 BC) set up trading links between Egypt and many countries, and sent people to places as far as modern-day Somalia for botanical treasures to plant in her gardens by the Nile.[3] Napoleon's first wife, Joséphine de Beauharnais (1763–1814), built up what was to be then the largest rose collection in the world, at the Château de Malmaison.[4] One of the first plant addicts, however, gardened in England. Mary Capel Somerset, Duchess of Beaufort (1630–1715), became a distinguished botanist. She married Henry Somerset in 1657 and when he inherited the Badminton estate in Gloucestershire she focused her attention on the gardens, developing a huge collection of plants, many of which were tender species that were grown in the gardens' extensive conservatories. She was personally involved in mounting herbarium specimens of many of her plants and became very knowledgeable about them. She wrote to her London neighbour Sir Hans Sloane, 'When I get into storys of plants I know not how to get out.'[5] The Dutch botanical artist Everard Kickius was hired to paint her collection and lived at Badminton for two years, from 1703 to 1705, while he did so.[6] Many of his paintings were designed as miniature landscapes with little regard for natural geographical or seasonal combinations of plants, but the plants themselves are portrayed with great accuracy. His beautiful illustration of the common snowdrop (*Galanthus nivalis*) is shown in an unlikely rocky landscape with a North American native thorn apple (*Datura stramonium*) and two sedum species.

The French journalist and novelist Jean-Baptiste Alphonse Karr is probably most famous for an epigram first used in the January 1849 issue of his journal *Les Guêpes* (The Wasps): 'Plus ça change, plus c'est la même chose' (The more it changes, the more it stays the same). In an introduction to J. J. Grandville's *Les Fleurs animées* Karr pointed out that 'We may love flowers in several ways. The naturalist flattens and dries them. He then inters them in a sort of cemetery called a herbarium and underneath them writes pompous epitaphs in a barbarous language.' Most gardeners, however, want to grow the object of their interest rather than flatten it, and some will go to great lengths to grow the plants they love in less than ideal conditions. An anonymous writer in the *Gardeners' Chronicle* of 1842 reported on his attempts to grow snowdrops in the West Indies, writing: 'I may state, however, that I was quite baffled in attempting to flower the Snowdrop; which may be accounted for by the impatient anxiety of the cultivator, with whom that interesting flower was always a favourite.'[7]

George Heath, who was described by his fellow writer Robert Buchanan as the Moorland Poet, included snowdrops in several of his poems, but it is perhaps from his diary that one can gain a greater appreciation of his love for the flowers. On 5 February 1869 he wrote:

> Fine and mild as April. Have been all about the fields, and my heart has been thrilled beyond measure by the appearance of several beautiful and only-just-peeping daisies. The hyacinths, too, are actually springing, and the celandine is out in leaf. How magnificent are the snowdrops! These flowers seem to

> my barren and often sadly yearning spirit like my own children – something I have a right to love and cherish.

Ellen Willmott (1858–1934) is probably as renowned today for her lavish spending habits as she is for her horticultural abilities. Graham Stuart Thomas, however, reported that 'she had beauty, intelligence, talent in many ways and the aptitude for complete absorption in whatever she was doing.'[8] Willmott joined the male-dominated Royal Horticultural Society in 1894 and within three years was working on its Narcissus committee. She raised many new daffodils and funded plant-collecting expeditions. In 1905 she became one of the first women to be elected a fellow of the Linnean Society of London. When in 1922 Willmott's sister Rose died of cancer, Willmott wrote forlornly to friends, 'now there is no one to send the first snowdrops to.'[9]

The Crimean War of 1853–6 is most famous today for the work of Florence Nightingale, and for the ill-fated Charge of the Light Brigade and other examples of the maladministration of the British Army.[10] In July 1853 Russia had occupied territories in the Crimean peninsula on the edge of the Black Sea, in the south of what is today Ukraine. Britain and France wanted to limit Russian expansion as the Ottoman Empire declined and had anticipated that their superior training and technology, such as explosive naval shells, would result in a short war. The conflict rumbled on, however, due to a lack of strategic planning. A year-long siege was required to take the naval base at Sebastopol. Four times as many British soldiers died of disease during the conflict as died in combat. The soldiers had to endure hard winters in the hills but as the snows melted in spring they would see snowdrops flowering, which may have reminded them of home.

At the end of the war many trophies were brought back to Britain. Three seventeenth-century church bells now in Arundel Castle, West Sussex, were taken from Sebastopol. Bulbs of the native snowdrops were rather easier to transport in a soldier's luggage, however, and many Crimean snowdrops made their way to Britain, where they soon made themselves at home. The *Illustrated London News* of 5 April 1856 carried a report on the Crimean snowdrop, *Galanthus plicatus*, pointing out its close similarity to *G. nivalis*, which was familiar to English gardeners, but emphasizing that the former can be distinguished easily by its pleated leaf margins. The writer thought the snowdrop a suitable emblem to offer consolation to the soldier far from home and hoped that the flower would 'burst through the snow-wreath like a resurrection from the dead to cheer his desolate heart, to whisper of coming succour, of brighter skies, of comfort and plenty, and glory in the field.' The article ended with the suggestion that the Crimean snowdrop may unhappily 'become a "memory flower", for many a dear friend who has gone down into a cold, cold, grave in the service of his country.'

The modern obsession with snowdrops could probably be said to date to 10 March 1891, when the first conference on snowdrops was held in London. It was organized by the Royal Horticultural Society and included lectures by James Allen, Frederick Burbidge and David Melville. Allen discussed the collection of snowdrops that he maintained at his home, Highfield House in Shepton Mallet, Somerset. He raised large numbers of seedlings and named over a hundred of them, although E. A. Bowles was later to comment that '[Allen] had such a keen eye for a slight difference that a great number of these seemed to other people very similar.'[11] Most of Allen's seedlings are thought to be no longer in cultivation. Highfield House was Grade II listed by English Heritage in 1974, but the garden was redeveloped when the house became the headquarters for Mendip District Council in 1981. The snowdrop that bears Allen's name, *Galanthus × allenii*, was named for him by John G. Baker in 1891. It is thought to have originated among some bulbs that Allen bought from an Austrian nurseryman by the name of Gusmus, and is probably of hybrid origin.[12]

Burbidge was curator of the Botanic Garden of Trinity College Dublin from 1879 until his death in 1905. He was a prolific garden writer of articles and books such as *Cultivated Plants: Their Propagation*

April 5, 1856. THE ILLUSTRATED LONDON NEWS 359

made at Constantinople, and much importunity for the aid of the English transport service, obtained an arrangement by which his army could be conveyed by instalments, at least, to the eastern shore of the Black Sea. Not knowing how reduced were the provisions of the heroic garrison at Kars, or how little chance there was of any direct relief being sent to them from Trebizond and Erzeroum, he proposed to effect a grand diversion in their favour; and, at the same time, change the whole aspect of the Asiatic war. It is probable that, had Kars been in the condition which he supposed, no better plan could have been selected. Proceeding first himself to Abasia, he fixed on Soukoum Kaleh as the fittest place for debarkation. He then returned to hasten the transport of his troops; but the fleets were then very busy, not only with the conveyance of supplies for the immense camps established on so many far-separated points, but with the transport hither and thither of large expeditionary detachments.

The most important of these went to the delta of the Boug and Dnieper, where, near their confluent estuaries, Otchakoff on the retreating northern, and Kinburn on the projecting southern bank, defended the approach from the west, to the dockyards of Nicolaieff, and of Kherson. On the 17th of October the garrison of Kinburn, 1500 strong, under Major-General Kokonovitch, surrendered to the Allies, who, having disembarked some troops to the north, bombarded the place both by sea and land. The French—still beforehand with us—here brought into action, for the first time in warfare, the new floating batteries invented by the Emperor Napoleon III. They succeeded admirably, the balls of the fortress, which mounted 174 guns, rebounding innocuously from their crab-like shells, at a range of only 800 yards. Otchakoff was then blown up and evacuated by the Russians.

Three weeks before this occurred the cavalry combat of Khougill, fifteen miles north-east of Eupatoria, between the Cossacks of General Korff and the French horse under General d'Allonville. On the very same day, the 29th of September, at the other side of the Euxine, General Mouravieff, having learnt the fall of Sebastopol, and wishing to expiate that dishonour, assaulted Kars with his whole infantry. The battle lasted eight hours, the Russian masses displaying the greatest stubbornness as they advanced under the well-arranged fire of the Turkish guns. At last, beaten everywhere, they were assailed in turn by the garrison which had scarcely yet been under fire; the confused and fatigued assailants could not make head against this sortie, and were driven to seek shelter under the protection of their numerous and powerful cavalry, leaving eight thousand dead in the trenches and near them.

It was at this time that Omer Pacha, gradually overcoming the difficulties of that deficiency in transports to which we have adverted, found himself at the head of about 15,000 troops in Abasia, a good many of these being his own trustworthy veterans. He had gradually edged them down towards Redoubt Kaleh, which he fixed on as his basis of operations. On the 30th Shemseral was secured; and, having driven in the Russian outposts from Sogdidi and endeavoured to open relations with Schamyl, while conciliating the Princess Dadian and the Christian population in his own neighbourhood, he moved inland in a south-eastern direction. But, owing to the difficulties of the country, which is an entanglement of woods interspersed with very rich but neglected farm lands, and owing still more to the precariousness and uncertainty of any supplies from the inhabitants, his progress was extremely slow. He did all that lay in his power, purchasing provisions in every direction, and organising, as his principal resource, a regular commissariat at Redoubt Kaleh. The river Phasis, which flows from the Caucasus to the Euxine, is navigable for nearly a hundred miles from the sea; and he had hoped to have availed himself of this channel for important manœuvres. His plan was this:—The first strong Russian post was at Kutais, where the great high road—by Gori, into Georgia, and down to Tiflis—would take his advancing columns over the celebrated Soorem Pass. Once master of Kutais, and with his communications well secured upon the Black Sea along his rear line, he hoped either to be able to defeat all the local Russian garrisons and posts between Soorem and the capital of the fertile province lying beyond and below it, or else to recall by the terror of his progress the army of General Mouravieff, then menacing Armenia, and beleaguering Kars. In either case a great blow would be struck, and the hard-pressed troops of General Williams relieved. Then, should it even prove too late to advance permanently that year beyond Mingrelia, he could at least strengthen himself in Kutais, make it his new centre for future operations, and call up, meantime, additional forces for the campaign of spring. General Mouravieff would then be pressed from the side of Armenia, where he was now acting offensively, and from the side of Imeretia, on which he would be thrown also upon the defensive. But it was already too late; and the Russian chief knew it. Well informed of the true state of the Kars garrison, he never disquieted himself, or in the slightest altered his plans, in consequence of Omer Pacha's diversion. Should the Muchir even beat the militia which now guarded the northern gorges of Georgia, he felt sure that it would all come to the same result. The season, the floods, scarcity, would compel the victor to retreat; much more would such become his necessity if, in the interim, he, General Mouravieff, should succeed in reducing Kars, and, while thus liberating his own army for an encounter with the Ottoman, should rob the latter of the chief motive which prompted this venturous advance by depriving it of its character *as a diversion*. Indeed, in such a contingency, the further Omer might have penetrated the worse would perhaps be his situation; since General Mouravieff, by not returning directly towards Tiflis (which would be rather better able than Kars had been to stand a siege in its turn and to hold any assailant in play), but, by moving diagonally, north-east by north, along the excellent Russian line from Alexandropol to Akhazik, would himself take Omer Pacha in flank and rear, shatter his line of communication, overwhelm his detached supports, and cut him off from the sea.

For these reasons, General Mouravieff tranquilly and steadily persisted in the blockade of Kars; and never for a moment showed any inclination to turn aside to face the Turkish invader. A month and seven days had now elapsed since the assault on Kars was repulsed so gloriously, when Omer Pacha at length brought his labouring columns through the miry woodlands as far as the Ingour. There he saw, for the first time, a regular stand prepared by the enemy, about twelve thousand strong, intrenched on the opposite bank, and commanding the passage by batteries. They were chiefly the Russian militia of Georgia, and were under the command of General Bragation-Mukransky. The Turks had some twenty thousand men. The stream was barely fordable in half-a-dozen places, by which the enemy's intrenchments could be turned. The Turks passed it, up to their armpits in water, holding their muskets aloft; our countrymen—Colonel Ballard, Captain Dymock, and others—showing a splendid example worthy of English officers. The engagement lasted five hours, when the Russians fled, leaving behind them sixty prisoners, five gun carriages and ammunition carts, and 400 killed. They appear to have carried off their wounded. Omer Pacha had 220 wounded and 68 killed. Pressing on the track of the fugitives he came up with them again before the end of November, within sight of Kutais, and obtained another advantage. But the floods had come; the Phasis had assumed the dimensions of a torrent; great forest trees were swept down the stream as if they were reeds—now engulfed out of sight in the eddies, now reappearing on the surface for a moment as they were borne away; the roads were impassable to artillery, and almost to infantry; the whole country was transformed into an alternation of morass and lagune; a day's march was the work of a week; the troops were broken up and islanded, as it were, into helpless detachments; the commissariat could not act; the supplies arrived with greater and greater irregularity, incertitude, and insufficiency from day to day; the whole army was suffering incredible hardships and privations; it was threatened with annihilation unless a retrograde movement were promptly made; and, finally, came the news that Kars had succumbed at last. The conquerors, therefore, retired, unpursued, and gradually straggled back to Redoubt Kaleh, where Omer Pacha soon succeeded in restoring their tone and refreshing their energies.

So ended the war of 1855 with Russia; for this was really its last incident, General Mouravieff having already dismantled the fortifications of Kars, and withdrawn the bulk of his forces to Gumri. It was on the 28th of November that General Williams at last surrendered to him the stubborn Armenian fortress. The heroic garrison had long been macerated by the failure of rations and by disease. Even their ammunition was expended. In another assault on the day of their surrender they would have had no means of firing half a dozen rounds from their guns, and they were completely past the power of personal resistance as a body, being unable to wield their weapons, and hardly able to stand erect. They had borne literally the fiercest extremities of famine. They were now a corps of spectres, with scarcely the strength to speak. Yet these men had furnished indiscriminately the sentinels who had mounted guard over the little pile of half-rotten farina which was to be doled out in a biscuit a day for each; and the trusty sentries never touched the food which was the sole remaining common stock. Under such circumstances it was that General Williams rode out with a flag of truce, and told Prince Mouravieff that he would surrender Kars provided all the courtesies and honours of war were conceded to the garrison. General Kmety and some few attendants had tried a different expedient—they stole out and cut their way through the leaguer on the only serviceable horses left.

Mouravieff listened with attention to General Williams, who threatened, if his various stipulations were not granted, to burst every gun and destroy every military trophy still extant in Kars. The Russian chief replied with chivalrous warmth and visible emotion as he looked at the emaciated hero, that all, all was granted, and that he was proud as an enemy to testify that General Williams and those under him had immortalised themselves. Nothing, in short, could surpass the nobility of sentiment displayed (both then and in the subsequent treatment of the prisoners) by Prince Mouravieff and the Russian army.

During the lull which winter now imposed upon hostilities Austria ascertained from the Western Powers what were the lowest conditions they would accept from Russia as the price of peace. These in principle amounted to the cession of all the pretensions which had provoked the war, and material guarantees against any renewal of such pretensions on the part of Russia. No more Greek protectorates; no fortresses on the Euxine or commanding it; no ships of war there; freedom of its waters to the commerce of all nations; no rebuilding of Bomarsund; the surrender of the Russian territory hitherto enthralling the mouths of the Danube; the dismantling of all arsenals on the great rivers within navigable distance of the Euxine; and all further provisions necessary to bring these general terms into practical effect.

Austria sent these propositions to St. Petersburg by Count Valentine Esterhazy on the 1st of the new year, giving Russia fourteen days to accept or reject, but not to modify them. The Czar sent counter propositions. Count Buol would not even hear them. At last, on the 15th the Czar telegraphed to Prince Gortschakoff at Vienna that the conditions were accepted without reserve. Paris was chosen as the scene of a Conference. Count Orloff and Baron Brunnow for Russia met the Representatives of France, England, Sardinia, Turkey, and Austria. An armistice was proclaimed till the 31st of March. Prussia was afterwards admitted by courtesy to ratify, but not to deliberate on, the essential preliminaries. At last, on the day before that when the armistice was to have closed, peace recommenced. The general conditions were all amicably settled: some of the smaller details were referred to the care of future executory commissions—one to adjust the Danubian cession of territory, another to define the frontier question of Armenia in Asia Minor. All the labours of the Congress were performed in less than a score of meetings; and, at two o'clock on Sunday, the 30th of March—just two years and two days after the declaration of war by the Western Powers—this great settlement of Europe and Asia was proclaimed. The instrument which had figured most largely in the prosecution of hostilities was that which announced their termination to the public, and at ten at night London heard and understood the expected salvos of artillery.

M. G. K.

CRIMEAN SNOWDROP.

While our brave soldiers have been fighting our battles in the Crimea the din and glory of war have not banished from their thoughts the arts of peace. Our conquerors do not, indeed, go forth like Alexander of old,

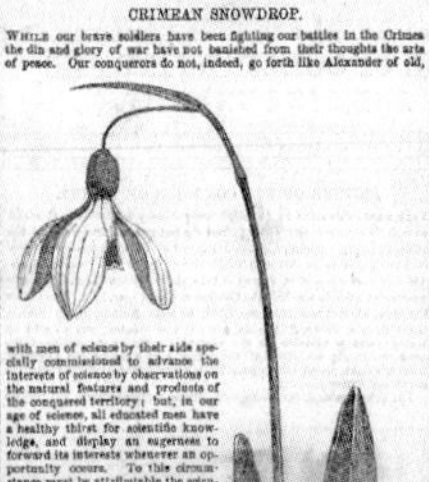

with men of science by their side specially commissioned to advance the interests of science by observations on the natural features and products of the conquered territory; but, in our age of science, all educated men have a healthy thirst for scientific knowledge, and display an eagerness to forward its interests whenever an opportunity occurs. To this circumstance must be attributable the scientific knowledge which has so suddenly been opened up to us respecting many features of the Crimean peninsula, which so long remained a *terra incognita* to the science of Western Europe. Its geology, its physical geography, its climate, its zoology, and its botany we were almost ignorant of when it was first proposed to reduce its forts. Now we know more or less of all—too much, indeed, from sad experience. But in addition to our knowledge we have acquired some useful material contributions to the arts of peace. Our soldiers have sent home seeds of trees new to our forests that may yet be grown into valuable timber; they have sent fruit-trees that promise greatly to enrich our orchards; and melon and other garden productions that have already rewarded our gardeners' care by the production of delicious fruit. Nor has the purely ornamental department of gardening been neglected; our flower gardens already boast of humble trophies that will bloom and fade and bloom again year after year, long after the siege of Sebastopol has gone down into the regions of remote history. One of these we have now the pleasure to introduce to our readers—the Crimean Snowdrop.

Emigrants, soldiers, and travellers, who go to distant lands, are prone to seek out wild flowers having some likeness to those of their native land, and to bestow upon them the names of early childhood, and thus link them with endearing associations. Thus the names of some of our common British wild flowers are applied to totally different plants in the Australian wilds. But in the case of the Crimean snowdrop there is no such fanciful resemblance to our English snowdrop. It is a true snowdrop, and is so strikingly like the common one that it was not till the middle of the sixteenth century that it was found to differ from it in any respect. The first mention of it is by Clusius, who states that a single bulb had been received from Constantinople, by way of Vienna, in 1592, but was subsequently lost. In *Hortus Kewensis* (1811) the plant is not even mentioned; but Dr. Fischer, the celebrated Russian botanist, had subsequently grown it in the Gorenki Garden, at Moscow, whence specimens were brought by Mr. Griffin to England.

At the monthly meeting of the Botanical Society of Edinburgh in February, 1856, Mr. McNab exhibited from the Botanic Garden plants in flower of the Crimean snowdrop, which had been raised from roots collected in the Crimea by Dr. Sinclair. The scientific history and characters of this plant may be briefly alluded to. It is the *Galanthus plicatus* of Marschall à Bieberstein (Pallas, Flor. Ind. Taur). It is mentioned by Clusius, in *Historia Rariorum Stirpium Pannoniæ*; and good figures and descriptions have been given in *Bot. Mag.*, t. 2162; and *Bot. Reg.*, 545. There is an early notice in *Bot. Hist.*, 1114.

The Crimean snowdrop is well distinguished from the common one by several essential characters, although its general appearance is so similar as to have caused it to be long overlooked as a distinct species. It is in every respect a larger and stouter plant than the English snowdrop; the outer divisions of the perianth (flower) are nearly twice the size, and it is as fragrant as the spring leucojum. The green markings on the inner divisions of the perianth are larger, and the flower-stalk is stouter, but sometimes scarcely so long as the common kind; indeed in one variety of *Galanthus plicatus* the spathe, which at first incloses the flowers, is longer than the flower stalk. But the leading features of distinction are found in the leaves of the *Galanthus plicatus*. These are very broad, more like the leaves of a narcissus than of a snowdrop, and they are furnished with two curious plaits, which run longitudinally from the base to the apex of the leaf near its margins.

Poets and writers on the language and poetry of flowers tell us that the English snowdrop is the emblem of "adventurous friendship" and of "consolation." We pretend not to be deeply versed in that mysterious department of philological learning called the emblematical language of flowers, and ought to be wary, therefore, in recommending innovations; but we may fairly anticipate the sympathy of our readers in recommending the extension of the same "sentiment" to the Crimean flower which is enjoyed by its English relative. If the "first pale blossom," "the early herald of the infant year," proves a "consolation" to us prosers and poets who peep at rough nature through the window-pane in earliest spring-time, ere the snows have gone—how much more eloquently would the fragile snowdrop speak of returning spring-time to the hearts of our noble soldiers while they were perishing in thousands under the rigour of a Crimean winter! We at home view the snowdrop, baffled by the ruthless storm, as the "herald of a brighter bloom." "Pleased we hail thee spotless blossom," for thou art rich in promise of green fields and flowery meads, singing birds, bright skies, and sylvan beauty! Thou bearest the welcome tidings of the coming of thy fair sister, the lily of the valley; and of her meet companions in the May wreath, fair narcissus by the mirrory waters, and the dancing daffodil of the woodland glade! But to the Crimean hero, bent down in disease and sorrow, far from home, and friends, and help—death and desolation all around him—no ray of hope to dispel the dark visions of his soul, thou, beneficent floweret! burst through the snow-wreath like a resurrection from the dead, to cheer his desolate heart, to whisper of coming succour, of brighter skies, of comfort and plenty, and glory in the field. Thou wert, indeed, "a flower of consolation" to him—a star of hope, "bright as a star when only one is shining in the sky." "God," saith Wordsworth,

> Made the flowers to beautify
> The earth and cheer man's careful mood,
> And he is happier who hath power
> To gather wisdom from a flower,
> And wake his heart in every hour
> To pleasant gratitude.

And this flower of the Crimea, the "Oriental Snowdrop," will in time to come oft realise that beautiful simile of Shelley, of—

> A pale flower by some sad maiden cherished,
> And fed with true love tears instead of dew.

It will become a "memory's flower," for many a dear friend who has gone down into a *cold*, *cold* grave in the service of his country.

It only remains for us to add that the kindness of Professor Balfour and Mr. M'Nab, of the Edinburgh Botanic Garden, has enabled us to give the accompanying drawing of this interesting plant, and that we are indebted to these gentlemen for some of the facts stated.

A Woman 100 Years Old.—There is at present living in Bancroft-road, King-street, Farnworth, a woman named Betty Pearson, who has attained the extraordinary age of 100 years. She was born at Bedford, near Leigh, on the 21st of December, 1755, and baptised at Culcheth-hall, near Warrington, there being no place of worship in that district at the time, and from which latter place her register of baptism has since been procured. Mrs. Pearson has been married twice. She was united to her first husband at the age of 24, and during his lifetime she had 11 children, he having now been dead 55 years. She had one child by her second husband, who also died 26 years ago. She has resided at Astley, in the same house, for the long period of forty years; but until very recently she resided at the White Horse in Leigh, serving in the capacity of "barmaid," and creating no small amount of curiosity among those who saw her while there. Her seventh child, a daughter, became the mother of 16 children, 10 of whom were born in the short period of 10 years. The daughter with whom she at present resides is the youngest of the 11 children, and is in her 60th year, having had 18 children, 12 of whom are now living. Mrs. Pearson has lived to see 35 grandchildren, 80 of her grandchildren's children, and 12 of their children, her great-great-grandchildren—in all 57 persons. She has been of very temperate habits, living on the plainest food, and during her early life was accustomed (using her own words) "to carding and spinning wool on her knee." She worked as a handloom weaver until 80 years of age. Her sight and other faculties are as good as those of most people of 70; having only lost two teeth, she is able to crack nuts with the greatest ease. She is in good health.—*Manchester Guardian*.

A Public Park in New York.—It is now definitively settled that New York city is to have one of the largest and finest parks in the world. It was feared for many years that this park never would be laid out. The chief obstacle to it was the natural construction of the island, which admitted of no place near the centre of population sufficiently large to be worthy of the name, since the island is so narrow that a park of large extent in any one place would cut the city in two. The expense, from the immense value of real estate, would be a heavy burden upon the municipality, and there was so little interest felt in the subject that only a few enthusiasts indulged in the dream. But the appropriation has been made. The park has been surveyed. The ground has been devoted to that purpose, and at an expense which is now known to reach some 5,000,000 dollars, with other expenses for adornment, which will probably extend to several millions more. The great work is now being put under way. The whole area of the park will be about 800 acres. It will stand nearly in the centre of Manhattan Island, on which New York stands, and over the whole of which its municipal authority extends. It will stretch through more than half of the breadth of the island from the Hudson to the East river in its widest part. It is wild and romantic in its scenery, and if it is laid out with taste will make altogether the most picturesque, wild, and romantic city drive there is within the limits of any great town in the world.—*Letter from New York*.

A Ghost Story.—A curious adventure has taken place in a district close at hand. The family of a peasant, who inhabits a solitary house in the fields, were celebrating the birth of a child. The happy father, in the excess of his joy, confided to the midwife the secret of his having saved up the sum of eighty dollars for the baptismal feast. A few days before the feast a figure made its appearance wrapped up in a bull's hide, with horns on its head, and, announcing itself as "the devil," demanded of the parents their new-born child. The father and mother groaned with distress, upon which the stranger consented to receive, instead of the child, a sum of eighty dollars, which were told out to him. He then inquired where they kept their provisions. They referred him to the loft. Whilst he was unhooking their sausages and pieces of bacon, a sportsman of the neighbourhood happened to look in, when he found the parents kneeling at their prayers. The peasant informed him that a terrible personage was in the house. The sportsman instantly mounted the staircase, crying out, "Who goes there?" A deep bass voice replied, "The devil!" The sportsman raised his gun to take aim, but at the noise the stranger called out, "For heaven's sake don't fire! I am N. N." It was the midwife's husband, and he is at this moment in prison.—*Posen Gazette*.

Value of Landed Property in Sydney.—As an evidence of the healthy state of money matters in this city, and in confirmation of the opinion so often expressed that property in good situations will always maintain its value, we may mention that at a Government land sale which took place on Tuesday, the 27th November, the Sydney Insurance Company purchased the corner allotment of Hunter-street and New-street at £101 per foot—being a price unheard of before in this city, the rate per acre being £70,000. Other lots sold at the same time, and in the same line of street, but in less advantageous parts of it, realising from £46 to £75 per foot, and the land disposed of yielding all overhead an average of £42,182 12s. per acre. When completed New Pitt-street will be the finest street in Sydney.

Large Sale of Negroes.—The *Washington Daily Union*, of the 12th March, has an auctioneer's advertisement, commencing—"On the first Monday in April next (the 7th), I will sell, at Cahawba, Dallas county, Ala., to the highest bidder, for cash, or Mobile 'acceptances,' at sixty or ninety days, bearing interest, one hundred and fifty likely negroes; also, fifty others, on longer time. All acclimated and experienced cotton-plantation hands. I will also sell at any time during this year at private sale (for cash or on time) two hundred other valuable negroes (in lots to suit purchasers), and 10,000 acres of the very best and most advantageously situated 'cane brake' and 'black land,' &c.

At the meeting of the Council of Revision of the Loiret last week, to decide on the exemptions from military service, the number of payments made for substitutes was three hundred and twenty, representing an amount of not less than 900,000f. This shews (says the *Constitutionnel*) that the new mode finds favour with the families of the above department.

Crimean snowdrop in the *Illustrated London News*, 5 April 1856.

and Improvement, published in 1877. He was described as having 'a pair of the richest sparkling brown eyes suggesting the happy monk endowed with a love of nature'.[13] In writing about Thomas Acton's garden at Kilmacurragh in County Wicklow, Burbidge described how after viewing such gardens 'one feels an exaltation of the mind, and a consciousness of it being a something more pleasant and satisfying than a jam tart-like garden of the carpet beds, or crowded flower show.'[14] He had worked as a plant collector for the Veitch nurseries in the late 1870s and is commemorated in the genus *Burbidgea* in the ginger family, having discovered plants of this genus while on an expedition in Borneo. His experiences on this expedition are recorded in his journal, which was published in 1880.[15] Burbidge appears to be the perfect example of that treasured species, the generous gardener. When a correspondent wrote asking for information concerning an autumn-flowering snowdrop that Burbidge grew, he wrote back enclosing the original bulb, only retaining an offset for himself; 'This great kindness to a perfect stranger was quite unexpected.'[16]

Burbidge's paper at the RHS conference reviewed snowdrop morphology, emphasizing the importance of looking at the leaves – not just the flowers – to help identify plants.[17] He also reviewed the majority of cultivars known at that time. He grew several selections of *G. reginae-olgae* at the Trinity College gardens, including some that had been collected in Greece by Professor John Pentland Mahaffy.[18] Mahaffy was born in Switzerland to an Irish family and became Provost of Trinity College Dublin in 1914 at the age of 75. Oscar Wilde wrote of him,

> I got my love of the Greek ideal and my intimate knowledge of the language at Trinity from Mahaffy and Tyrrell; they were Trinity to me; Mahaffy was especially valuable to me at that time. Though not so good a scholar as Tyrrell, he had been in Greece, had lived there and saturated himself with Greek thought and Greek feeling.[19]

Two of Mahaffy's autumn-flowering selections of G. *reginae-olgae* were named after his daughters, Elsie and Rachel. G. 'Elsae' was collected on Mount Athos in Greece around 1888. It was reported to be in bud in Dublin in December but had been known to open earlier. G. 'Rachelae', which arose from a single bulb collected on Mount Hymettus in 1884, had flowers like those of G. *nivalis* but broader leaves. A report on snowdrops in *Garden and Forest* of 22 February 1893 stated that 'Rachelae' and 'Elsae' were in cultivation 'but are so rare as to be almost priceless'. It was thought at the time that there were just 91 authenticated bulbs of 'Rachelae' in existence. An increase from one bulb to 91 in nine years might be considered a satisfactory rate to many gardeners, but it was not enough to ensure their continued survival, and 'Rachelae' is now considered to be lost to cultivation.

David Melville (d. 1924) was gardener to the dukes and duchesses of Sutherland, first in Staffordshire then as head gardener at Dunrobin Castle in Sutherland, in the Highlands of Scotland, a position that he held for 46 years. The snowdrop first exhibited as 'Dunrobin Seedling' in 1878 was later named G. *nivalis* 'Melvillei' to commemorate Melville. It flowers rather earlier than normal forms of the species. The original G. *nivalis* 'Poculiformis' was found at Dunrobin and discussed in Melville's lecture at the RHS conference. In this form the normal green-marked inner segments are absent and are replaced by an extra set of outer petals, so that there are two whorls of white petals. The word 'poculiformis' means 'shaped like a little cup' and was first used in 1880 by the Reverend Henry Harpur-Crewe to describe Melville's find. Similar plants do occur from time to time in large snowdrop populations. 'Serotinus', another Dunrobin seedling, was a very small form.

Harpur-Crewe (1828–1883), a clergyman and naturalist, was also responsible for naming G. *nivalis* var. *sandersii*. This had been found by a Mr Sanders of Newnham in Cambridge in a garden near Belford in Northumberland and had a yellow ovary and yellow markings on the inner segments. Sanders sent it to Harpur-Crewe, who corresponded with many snowdrop growers of the Victorian era and distributed the snowdrop among his acquaintances.

Many other members of the clergy were interested in the natural world, perhaps most famously the parson-naturalist Gilbert White, who wrote *The Natural History and Antiquities of Selborne* (1789). White noted in his journal in February 1770, 'Yellow-hammer sings and bees gather on the snowdrops.' Bitton, the Gloucestershire parsonage of Henry Nicholson Ellacombe, was described by him in his books *In a Gloucestershire Garden* (1895) and *In My Vicarage Garden* (1905). They give a month by-month guide to the plants Ellacombe grew and to his insights into gardening. He exchanged plants and seeds with the Royal Botanic Gardens at Kew and many other gardens across Europe. One of the highlights of his garden was a display of the snowdrop G. 'Atkinsii'. Bowles wrote, 'Canon Ellacombe gave me this snowdrop and quite half of my garden treasures besides, and it is one of the floral treats of the year to see it in January growing over a foot high under the south wall at Bitton.'[20] At the RHS Snowdrop Conference, it was reported that James Atkins of Painswick had received the snowdrop from a friend in the 1860s and had given it to Ellacombe, who had distributed it widely.

The designer and critic William Morris warned, 'Be very shy of double flowers . . . Don't be swindled out of that wonder of beauty, a single snowdrop; there is no gain and plenty of loss in the double one.'[21] There are many people, however, who while appreciating the simple beauty of a single snowdrop also admire the ballerina-like double snowdrops, with their dainty tutus of inner petals. The best-known hybrid double snowdrops are those called the Greatorex Doubles, after the man who raised them, Heyrick Anthony Greatorex. Greatorex (1884–1954) was born in Brixton in London and sent to school in Derbyshire. Greatorex married Janette Elizabeth Tillett in 1915 and during the First World War was wounded in action at Lagnicourt and awarded the Victory Medal and the British Medal. From 1917 until 1952 Greatorex and Janette lived at Witton, about two miles from Brundall in Norfolk, in a small house set in about six acres of land. He grew a variety of interesting plants there, including snowdrops and colchicums.

Greatorex is thought to have been the first person to deliberately create hybrid snowdrops, transferring pollen from the double form of the common snowdrop (*Galanthus nivalis* 'Flore Pleno') to the flowers of *G. plicatus*.[22] *G. nivalis* 'Flore Pleno' does not set seed but can produce fertile pollen. Doubleness appears to be a dominant trait and many of the seedlings also seem to have inherited the vigour of the pollen parent. Fourteen of the resulting plants were selected and named by Greatorex himself. The majority of these were named after female characters from the works of Shakespeare. 'Nerissa', 'Hippolyta' and 'Lavinia' are known to have been released before 1948 because they feature in drawings made by E. A. Bowles that are now in the Lindley Library in London. Dates for Greatorex's other introductions are uncertain and he did not publish detailed descriptions of the cultivars.

Galanthus 'Nerissa' (Portia's waiting maid in *The Merchant of Venice*) is shorter than some of the other Greatorex doubles and has pretty ballerina-style flowers with variable markings on the inner segments. 'Hippolyta', named for the Amazon queen in *A Midsummer Night's Dream*, is one of the shortest cultivars, with a very neat skirt of inner petals. 'Ophelia', who loved Hamlet, is the most widely grown form, having been first marketed in 1959. It has pretty rounded flowers which usually have aberrant segments. 'Jaquenetta', named for a country maid in *Love's Labour's Lost*, is taller-growing and has large green marks on the inner segments. Sometimes the outer segments are also marked in green at the tips.

'Desdemona' was named after the heroine of *Othello*; it is one of larger clones and sometimes shows green tips to the outer segments. Its bulbs frequently produce a third leaf on the shoot. 'Titania', Queen of the Fairies in *A Midsummer Night's Dream*, is an excellent form with very regular flowers. 'Lavinia', after the daughter of Titus in *Titus Andronicus*, is one of the taller Greatorex doubles. The marks on its inner segments tend to bleed towards the base at the sides.

'Cordelia', after the youngest of King Lear's daughters, is a taller-growing clone with flowers like a fuller version of 'Lavinia'.

'Dionysus' has a tendency to sometimes produce single rather than double flowers. The Greek god of fertility and wine is not specifically mentioned in Shakespeare except in his Roman incarnation Bacchus; however, there is a Dionyza in *Pericles, Prince of Tyre*, which is attributed at least in part to Shakespeare. It is uncertain whether Greatorex intended this cultivar to be named for the character in *Pericles* or if he was planning a second series named after Greek gods, given that he sent to Frank Waley of Hertfordshire a bulb named 'Poseidon', after the Greek god of the sea. Two more Greatorex doubles, 'White Swan' and 'Jenny Wren', were listed by Stern in 1956.[23]

Shakespeare wrote in *The Merchant of Venice*, 'It is a wise father that knows his own child' (Act II, Scene 2). For snowdrop growers it may be difficult to know exactly which cultivars they are growing. Many of the Greatorex plants look very similar to each other and they have become much confused in cultivation. This is complicated by the inherent variability of these snowdrops: heights of the plants can vary with different cultivation conditions and the number of inner segments and their markings are not totally stable. Without having a definite provenance it would be a bold person indeed who would identify a particular plant with any degree of certainty. Some commercial suppliers just list the bulbs as 'mixed Greatorex doubles'.

Greatorex's garden was subject to a purchase order after his wife died in 1971. It is now in the care of Blofield and District Conservation Group and is known locally as Snowdrop Acre.[24] As well as the doubles, a variety of different snowdrops still grow on the site, including colonies of *G. woronowii* and *G. ikariae*.

One of the most lovely snowdrop species is *G. elwesii*, which was named after Henry John Elwes (1846–1922). Elwes was a larger-than-life character with an intense and wide-ranging interest in the natural world encompassing big game, insects, birds and botany. He published the folio-sized *Monograph of the Genus Lilium* in 1880 and the 1906 book *Trees of Great Britain and Ireland*, co-written with Augustine Henry,

A Greatorex double-flowered snowdrop, *Galanthus* 'Cordelia'.

described every species of tree then grown outdoors in the British Isles. Elwes was one of the first people to receive the Victoria Medal of Honour from the Royal Horticultural Society, which he did in 1897.[25] On a visit to Turkey in early April 1874, Elwes came across a 'fine large snowdrop' while in the mountains near Smyrna (modern Izmir). Before leaving Turkey he arranged for bulbs to be collected later in the year. The snowdrop was subsequently named *G. elwesii* by Joseph Hooker and was illustrated by W. H. Fitch in *Curtis's Botanical Magazine* in 1875. Somewhat ironically, the original plants collected by Elwes, which were used as the type specimen for *G. elwesii*, were later found to represent a broad-leaved form of *G. gracilis*.[26] The name *G. elwesii* was retained for a different type specimen.[27]

Elwes grew *G. elwesii* in his garden at Miserden in Gloucestershire. In 1891 he inherited the family estate at Colesbourne Park, near Cheltenham, where he developed a notable arboretum and a comprehensive collection of bulbous plants. The garden at Colesbourne was neglected after Elwes's death, but many of the snowdrops naturalized and in recent years the collection has been developed by the current

Sir Henry Elwes and his wife, Carolyn. Carolyn Elwes swapped bulbs with many other enthusiasts, building up a varied collection which is today regarded as one of the finest in Britain. The snowdrop collection at Colesbourne Park was opened to the public for the first Snowdrop Gala in February 1995 and attracted two hundred people. Interest has developed greatly since that time and the garden open days now attract thousands of visitors.

The Elphinstone name is famous in Aberdeen due to its connection with the founder of the university there; to the galanthophile, however, the name recalls a very special snowdrop. Sir Graeme Elphinstone was born in 1841 and married Margaret Anne Alice Fairlie on 5 January 1875; they had twin daughters, Mary and Constance, born in 1876. Mary died in 1919 but Constance lived until 1960. The family spent a lot of time abroad and owned a coffee plantation in Ceylon (now Sri Lanka). When troubled with disease in the coffee plants Elphinstone was one of the first plantation owners to move his enterprise to the Malaysian state of Perak. His estate of about 1,000 acres in the Kuala Kanga district successfully produced Arabica coffee.[28] He collected plants on the Taiping Hills near Perak and at the Gapis Pass; specimens of these are now in the Herbarium of the Singapore Botanic Gardens.[29] A letter from Benjamin Disraeli to Queen Victoria of 13 April 1858 reported on a speech made by Lord Elphinstone on the subject of the Straits Settlements, describing Elphinstone as a 'master of the subject' and praising his 'vindication of the Convict population of Sincapore [*sic*]'.[30] The Straits Settlements was a former British Crown colony on the Strait of Malacca, and it was there that Elphinstone died in May 1900 at the age of 58.

Elphinstone succeeded to the title of 4th Baronet Dalrymple-Horn-Elphinstone in February 1887, and found the snowdrop now known as *Galanthus nivalis* 'Lady Elphinstone' in 1890 when walking in the grounds of Heawood Hall, his Cheshire estate. The snowdrop is obviously closely related to *G. nivalis* 'Flore Pleno' in that it is a double with irregular inner petals, but instead of the usual green 'Lady Elphinstone' has markings of a soft apricot-yellow. It does have the somewhat disconcerting habit of reverting to green markings if moved, and snowdrop suppliers are used to people complaining that they have been sold the wrong cultivar. The intensity of colour of the markings can vary from year to year even in established clumps. Elphinstone gave bulbs of the snowdrop to the galanthophile Samuel Arnott, who distributed it, and it is now well-established in cultivation. 'Lady Elphinstone' was referred to by Graham Stuart Thomas in his book *Colour in the Winter Garden* as 'Countess of Elphinstone'.

Another Lady Elphinstone, Lady Mary Frances Elphinstone, was the elder sister of Elizabeth Bowes-Lyon, the Queen Mother. Her book *Botany for Children* (1936) describes the flowers found in Britain, arranged in their families rather than by season of flowering, as was more usual in books for beginners written at that time. It was illustrated with line drawings by the author and colour prints showing the plants in a natural setting with details of their roots and fruits by Amy Webb. It is written in a style which may well cause some derision nowadays, with three children being instructed in the art of plant identification by Dame Nature and a young woman-nymph called Flora. Flora gets the children playing 'follow my leader' and takes them to some snowdrops, where she talks about the spathe that protects the flower in bud and how the flower has an inferior ovary. The book was reprinted under the title *Flowers and their Families* in 1946. Mary Elphinstone is commemorated not with a snowdrop but with an apricot-coloured hybrid tea rose, 'Lady Elphinstone', which was introduced in 1921.[31]

Edward Bertram Anderson was a research chemist by profession but also a plantsman and horticultural writer and one of the founder members of the Alpine Garden Society. His books, including *Gardening on Chalk and Limestone* (1965) and *Seven Gardens; or, Sixty Years of Gardening* (1973), are considered to be classic horticultural texts. Anderson made the observation that snowdrops such as 'Merlin', with entirely green

Galanthus nivalis 'Lady Elphinstone', a double form with apricot-coloured markings in some seasons.

inner petals, can occur from self-sown seedlings, as had happened in his garden. The popular snowdrop G. 'Mighty Atom' originated in Anderson's Gloucestershire garden. As its name suggests, it has a very large flower, although the stem itself is not particularly tall. Anderson considered the ordinary G. *nivalis* 'too invasive for the flower beds'.[32]

The plantswoman and writer Beth Chatto, who has a renowned garden in Essex, claimed in her book *The Damp Garden* (1982), 'I am not yet a snowdrop fanatic, but I do enjoy finding really distinctive ones.' The word 'yet' is noteworthy in that statement, as galanthophilia often first presents as a mild interest in snowdrops. Chatto splits her snowdrop clumps immediately after flowering and sometimes even when they are in full flower. 'They do not resent this but flower the next year most rewardingly. To wait until a plant has died down often means I forget it until too late.'[33]

Chatto was a friend of the artist and gardener Sir Cedric Morris (1889–1982). In 1937 Morris and fellow painter Arthur Lett-Haines (1894–1978) opened the East Anglian School of Painting and Drawing, which attracted many pupils including Lucian Freud and Maggi Hambling. Morris was as interested in horticulture as he was in painting and he raised thousands of iris seedlings, many of which were named and are still in cultivation. Each June he held iris parties at his garden, Benton End in Hadleigh, Suffolk, which were attended by many famous gardeners including Vita Sackville-West.[34] There is a very lovely miniature daffodil named after him, *Narcissus* 'Cedric Morris', which is particularly treasured as it flowers in the middle of winter. The snowdrops *Galanthus* 'Benton Magnet' and 'Cedric's Prolific' both originated at Morris's garden.

Of course, snowdrops do not need to be new cultivars to be greeted with enthusiasm. The prolific author and playwright Beverley Nichols, remembered today mostly for his gardening books, which were written with humour and not insignificant amounts of irony, wrote of the excitement of seeing what plants come up in a garden that one newly owns:

> Of all the treasure hunts in which men have ever engaged, this must surely be the most enthralling . . . to wander out on a February morning, in an old garden which is all your own and yet is still a mystery, and to prowl about under the beech trees, gently raking away a layer of frozen leaves in the hope of finding a cluster of snowdrops.[35]

Nichols was particularly fond of winter flowers but warned that 'the only way in which we can appreciate the beauty of snowdrops is by going out into the garden, lying flat on our backs in the mud and gazing up at them from below.'[36] He did, however, suggest that to overcome this problem you could employ the 'Nichols' Snowdrop Invention', which uses a sheet of mirror in order to display cut snowdrops indoors.

Nichols recommended the species *G. elwesii*, noting that 'It is, of course, a more expensive bulb, but you would be expensive too, if you looked like that.'[37] He reported that a sizeable proportion of his income went to bulb suppliers, 'But I would rather be made bankrupt by a bulb merchant than by a chorus girl.'[38] Today the potential for bankruptcy following injudicious snowdrop purchasing is probably even higher than in Nichols's time, since there are many more specialist suppliers around.

The first specialist snowdrop supplier was the Giant Snowdrop Company, which operated from 1953 to 1968 following a highly acclaimed display of snowdrops staged at the RHS winter show in 1951. It was run by Brigadier Leonard Mathias and his wife Winifrede, who bought Hyde Lodge in Chalford, Gloucestershire, in 1947.[39] Hyde Lodge had previously been the home of Walter Butt (1880–1953), who moved to Bales Mead in Porlock, Somerset, in 1940. The garden at Hyde Lodge became somewhat neglected after Butt left but Mathias found that it was full of snowdrops, many of which had come from the Colesbourne collection. In later years E. B. Anderson took over Butt's garden at Bales Mead, and found there a soft lavender blue form of the early-flowering Algerian iris, which he named *Iris unguicularis* 'Walter Butt'. Leonard and Winifrede Mathias were helped in their snowdrop enterprise by their gardener Herbert Ransom, who had originally been their chauffeur and handy-man.[40] He is commemorated by the snowdrop cultivar 'Ransom', which is thought to be a hybrid between G. 'S. Arnott' and G. *plicatus*.

The Giant Snowdrop Company introduced the practice of selling snowdrops 'in the green' during the growing season. This is a somewhat controversial practice because it does cause a degree of root damage, but if the bulbs are dispatched and replanted quickly it can be a very successful method and reduces the risk of bulbs drying out in storage (which is a problem with snowdrops). At their peak, the company listed 26 varieties and were selling in the region of 25,000 to 30,000 bulbs each year.[41] They produced a comprehensive illustrated catalogue of their plants. The couple's daughter Armine has a tall and reliable snowdrop named after her.

North Green Snowdrops was established in 1984 by the artist John Morley (b. 1942).[42] His beautiful, contemplative paintings often feature plants grown in his garden at North Green Only in Suffolk. Morley's annual snowdrop catalogue is a sought-after source of information and often features a snowdrop painting by him on the cover. One of Morley's most treasured snowdrop introductions is 'Three Ships', which was found growing in leaf litter under a cork oak in the former garden of Henham Hall in Suffolk. It was so-named because it 'never fails to "come sailing by on Christmas day"'.[43]

Dr Ronald Mackenzie set up The Snowdrop Company in 1991 in response to his frustration at reading about bulbs which he was then unable to source. He obtained much of his original stock from Herbert Ransom at Hyde Lodge. Mackenzie was one of the first snowdrop suppliers to twin-scale bulbs, a method that involves cutting a bulb into small sections, each with a piece of basal plate and at least two scales to nourish the growing bulblet. One bulb can produce 32 or more offsets, and these often flower within three or four years, enabling lots of snowdrops to be raised quickly. Hardy Amies (1909–2003), the Savile Row fashion designer, loved snowdrops and was a customer of The Snowdrop Company, although he tended to grumble about the prices charged.[44] One of the special snowdrops that Mackenzie introduced into commerce was *G. elwesii* 'Godfrey Owen', a double in the true meaning of the word: it has six outer segments surrounding six neat inners. It was a chance discovery found in the Shropshire garden of snowdrop enthusiast Margaret Owen and named after her late husband.[45] Mackenzie regularly won awards for his snowdrop displays at Royal Horticultural Society shows and in 2003 was presented with the Alpine Garden Society Kath Dryden Award in recognition of his meritorious work and expertise on the genus *Galanthus*.

Many gardens open to the public when their snowdrops are in bloom and it is often possible to purchase snowdrops as a memento of the visit. Anglesey Abbey near Cambridge in the east of England has a renowned snowdrop garden on a former monastic site. Urban Huttleston Rogers Broughton, the first Baron Fairhaven, bought the property in 1926 and had the grounds laid out in an eighteenth-century style. On his death Anglesey Abbey was left to the National Trust. The special snowdrops on the site first emerged in 1970 when staff were clearing thousands of diseased elm trees from the estate. Surviving plants from discarded snowdrops were found still growing in a Victorian-era garden refuse area. Fifteen varieties were found and formed the basis of the Abbey's collection, but this grew rapidly when Richard Ayres, head gardener at the time, began swapping bulbs with other collectors.[46] There are now over 240 varieties of snowdrops grown there, many of which have naturalized throughout the grounds. The snowdrop 'Anglesey Abbey', which was discovered there, was first thought to be an example of the species *G. lagodechianus*, but more recent study indicates that it is an unusual variant of *G. nivalis*.[47] It is a vigorous plant with flat glossy green leaves; the flower shape and markings vary but blooms are often poculiform (cup-shaped).

Many other cultivars have arisen at Anglesey Abbey, including 'Anglesey Adder', which was named for the plant's split spathe which resembles an adder's tongue, and the highly sought-after 'Anglesey

Orange Tip', which caused something of a sensation when it was first displayed in 2010. This is an *elwesii* cultivar with pale apricot-flushed segments that hold their colour well. The inner segments have a broad inverted green 'U' at the apex. 'Moses Basket' is another unusual *G. elwesii* cultivar with a distinct form.

The annual Snowdrop Walks at Rode Hall in Cheshire are very popular with galanthophiles and the general public alike. Rode Estate has been in the ownership of the Wilbraham family since 1669. The snowdrops were originally introduced to the estate by Sibella Egerton at the time of her marriage to Randle Wilbraham in 1833. Extensive work has been carried out in recent years to restore the gardens to their nineteenth-century magnificence and over fifty varieties of snowdrop can be seen planted throughout a mile-long walk. A new feature in 2013 was a theatre for displaying pots of snowdrops in the manner of the auricula theatres that were popular among florists of the eighteenth and nineteenth centuries.

The Chelsea Physic Garden in London also has a snowdrop theatre. It is used to display many different varieties during the special snowdrop days that attract snowdrop lovers for walks, talks and shopping opportunities. One of the best places to see snowdrops in Britain is at the Painswick Rococo Garden in Gloucestershire, where paths wind through carpets of naturalized snowdrops under beech and other trees. The snowdrop grove there is thought to date to the mid-1800s and traditionally the garden was open to Painswick residents for one Sunday each year. The villagers were each permitted to pick a posy of six snowdrops. Unfortunately, since first opening to the general public in 1984 picking has had to be discouraged due to the huge number of visitors received. Nevertheless, a walk through the garden on a sunny winter day remains a truly unforgettable experience.

Mr Farebrother says in George Eliot's *Middlemarch*, 'We collectors feel an interest in every new man till he has seen all we have to show him.' Plant collections are not, however, necessarily just repositories of specimens kept to please the obsessive plantsperson. There are often valid scientific reasons for keeping a genetic diversity of species. Different populations of plants may be important for their medicinal or other properties. The use of galantamine as a treatment for Alzheimer's disease was discussed in the previous chapter and studies into other potential benefits of snowdrops are ongoing. There is a great deal of current research looking at the bioactive properties of snowdrops. Analysis of 25 populations of *Galanthus elwesii* and seven of *G. nivalis* collected from different locations in Bulgaria showed considerable variation in the profiles of alkaloids found in geographical variants. Adjacent populations show similar profiles, and experiments in which the snowdrops were transplanted imply that there were genetic rather than environmental causes for the different levels of alkaloids found.[48]

Widespread concerns about the adverse effects of pesticides on bee populations has led to research into the development of insecticides that will target pests while posing negligible risks to beneficial organisms. One such strategy is looking at linking insect-specific spider venom with the lectin, a carbohydrate-binding protein, from snowdrops.[49]

In the *Gardeners' Chronicle* of 12 March 1842 there was a report on the medicinal properties of members of the Amaryllidaceae:

> Hooping cough is cured by the extract of Narciss bulbs; Snowdrops, Snowflakes, and Daffodils are good emetics; the Sea Pancratium emulates the Squill in its utility for coughs, and all of them form excellent poultices for troublesome tumours. So true it is in the natural as well as the moral world that the bitterest things are often the most conducive to our welfare and that the danger of to-day is the salvation of tomorrow.[50]

Although there was apparent faith in the worth of these natural products, garden writers also knew of the dangers, and reported on a child who had eaten a daffodil bulb and had become very sick and delirious. The article warns: 'Be careful then, ye mothers, how you let your children play with daffodils . . . Even the Snowdrop and

Snowflake, for all their demure looks, are of the same vile race and are no more to be trusted than the Narciss.'

'A fool, you know, is a man who never tried an experiment in his life.' Erasmus Darwin (1731–1802), the English physician, naturalist and grandfather of Charles Darwin, was no fool but a great thinker – and he certainly carried out many experiments. Johann Gottlieb Gleditsch (1714–1786), a physician and director of the Berlin Botanical Gardens, was reported to have obtained 'an excellent starch' from snowdrop bulbs.[51] Darwin tried a similar experiment:

> Some snowdrop-roots taken up in winter, and boiled, had the insipid mucilaginous taste of the Orchis, and, if cured in the same manner, would probably make as good salep . . . Some roots of Crocus, which I boiled, had a disagreeable flavour.[52]

He also suggested damaging the bulbs or growing them tightly together in garden pots in order to stimulate the production of more seed, as a mode of mass-producing the bulbs for profit. While the latter idea may be worth trying, today boiling and eating snowdrop bulbs is not recommended due to the potentially toxic chemicals that they contain. Darwin was a poet as well as a scientist and included verses in his botanical theses:

> How Snow-drops cold, and blue-eyed Harebels blend
> Their tender tears, as o'er the stream they bend.

Some snowdrops, including *Galanthus nivalis*, *G. cilicicus* and the cultivar 'Scharlockii', have been found to contain a chromogen which when heated produces an azure-blue pigment known as galanthus blue.[53] The chemical is absent from *G. elwesii* and related genera including *Narcissus* and *Leucojum*.

Snowdrop Conservation

The main threats to wild snowdrop populations are the destruction of habitat and the over-collection of bulbs for the horticultural trade. The trade in wild-collected bulbous plants from Turkey dates back at least to the sixteenth century, when interest in exotic plants such as tulips and fritillaries was rising. Today snowdrops are the most heavily traded of all wild-collected bulbs, with G. *elwesii* the most common wild-collected species in commerce.[54] The Convention on International Trade in Endangered Species of Wild Fauna and Flora (CITES) is an international agreement between governments which aims to ensure that international trade in specimens of wild animals and plants does not threaten their survival.[55] The treaty first came into force on 1 July 1975. All snowdrop species were included in the CITES Appendix II in 1989. This applies to plants which are not under imminent threat of extinction but where the trade in them is monitored to ensure wild populations are not endangered. Snowdrops, whether live plants, dormant bulbs or indeed dead plants, cannot be moved between countries (except within European Union member states) without a valid CITES permit.

The trade of three species, G. *nivalis*, G. *elwesii* and G. *woronowii*, is limited for wild-collected bulbs. Individual countries can add further restrictions on trade. In 1986 Turkish legislation set up specific quotas for the export of G. *elwesii* to ensure sustainable trade. Collection sites must be allowed to regenerate for three years in Turkey before harvesting is permitted again.

Of course, where there is a law there will be somebody trying to find a way around it, and there are reports of farmers digging up bulbs of G. *woronowii* in Turkey and replanting them in Georgia to be bulked up for sale.[56] The Dumlugöze snowdrop project in Turkey aims to farm G. *elwesii* for export to provide valuable incomes for local people and to relieve the pressure on wild populations of snowdrops.[57] Under the Indigenous Propagation Project supported by the World Wide Fund for Nature (WWF) in Turkey, villagers receive money to grow snowdrop bulbs as a crop and are taught cultivation and propagation techniques.

In Britain there are currently three National Plant Collections of *Galanthus*, held under the auspices of the charity Plant Heritage, which aims to ensure the conservation of cultivated plants.[58] The collection holders are private individuals whose collections have been accredited by Plant Heritage. The snowdrop collection of Margaret and David MacLennan has over 600 varieties. Many of their snowdrops grow in neat raised beds, protected against narcissus fly by a fine mesh. Steve Owen in Leighton Buzzard in Bedfordshire has a collection that includes some 900 different species and cultivars. Owen combines his snowdrops with a fascinating range of other winter-flowering plants, such as crocuses and daphnes. The largest collection in terms of acreage is that of Catherine Erskine at Cambo in Fife. The Cambo snowdrop collection of around 350 taxa grows in 70 acres of private gardens and woodland. The Cambo Estate works with other Scottish gardens to run an annual snowdrop festival with special events including a 'Snowdrops by Starlight' experience in the Cambo Gardens.

Collections of snowdrops are at risk of various pests and diseases. A grey mould specific to snowdrops, *Botrytis galanthina*, can attack plants, particularly in mild, damp conditions. White fungal growth initially attacks the stem at soil level, before the rot spreads down so that the bulb scales go brown and the bulb becomes soft. The fungus' hard black food reserves or sclerotia found on the outermost scales and on shoots can survive for at least a year in the soil to reinfect developing shoots in subsequent years. Some species, such as *G. plicatus*, seem to be more resistant to infection.

Stagonospora curtisii is a fungal infection more commonly seen on *Hippeastrum* in the Amaryllidaceae family. It produces a red-brown discolouration at the top of the outer bulb scale and spreads downwards causing the bulbs to dessicate, and is by far the most devastating snowdrop disease for collectors.

Various stem nematode roundworms can attach themselves to the bulbs, causing distorted and stunted growth and browning of the bulbs. The narcissus fly (*Merodon equestris*) is not fussy about food and will lay its eggs on snowdrops as well as daffodils. It is the emerging maggots rather than the adult flies that cause the problem, as they can damage or destroy the bulb by eating it. Various other ground-dwelling pests sometimes attack snowdrop bulbs, for example the larvae of swift moths such as *Hepialus humuli* or *H. lupulinus*. In Holland the swift moth is known as the 'lettuce root-driller', as it was first found in lettuce crops in the early 1900s. It causes significant problems in the Dutch cut flower industry. The adult moths spend daylight hours concealed in vegetation and fly mostly on warm summer nights. The female deposits her eggs during flight; on hatching, the larvae burrow into the ground, where they feed on plant roots. The larva is an ugly white grub with a brown head. Some species may take two years to complete their larval development, so the larvae may be found in the soil at any time of year. Chemical control is difficult as the older larvae are tolerant of pesticides.

It is not only insect pests and diseases that snowdrop collectors have to be wary of. The high prices paid for individual snowdrops attract a lot of attention and unfortunately can lead to criminal behaviour. *Galanthus* rustling is a surprisingly big business; in January 2000 two men were imprisoned following conviction for stealing some 300,000 snowdrop bulbs, thought to be worth £60,000 (around $38,000).[59] The first yellow-marked form to be found of the prized species *G. elwesii* was 'Carolyn Elwes'. This appeared as a seedling at Colesbourne Park and was first noticed in 1983. It has soft lime-yellow inner segment markings. The tips of the leaves and spathes are also often yellow. It is a particularly desirable cultivar and, following the Colesbourne snowdrop open days in 1997, the original clump was stolen. The theft was widely publicized and because the cultivar is so distinctive it could not easily be sold on. Sir Henry Elwes compared the theft of the bulbs with stealing a painting by Monet.[60] No trace of the bulbs has ever been found, so it is not known whether the bulbs are gloated over, growing in a hidden corner of the criminal's garden, or whether they were thought to be too hot to handle

Galanthus elwesii 'Carolyn Elwes', with Carolyn Elwes behind.

and ended up on a compost heap. Fortunately, the head gardener at Colesbourne had saved a reserve of some half a dozen bulbs, and so this beautiful plant has not been lost to cultivation.

A tall, yellow-flowered form of *G. woronowii* was noticed in 2005 by Elizabeth Harrison while she was looking at plants in her Perthshire garden. Nurseryman Ian Christie realized its potential and worked up a stock of the plant, which he has named 'Elizabeth Harrison' after the finder. A single bulb was sold on an online auction site to the Ipswich-based seed company Thompson & Morgan for £725 ($1,100) in 2012, the top price then paid for a snowdrop bulb. The company hopes to be able to go into commercial production of the bulb via tissue culture. North Green Snowdrops had also been building up a stock for sale but their bulbs were stolen, and they also lost their entire stock of G. 'Trimmer'. Many owners of valuable bulbs today install security cameras and other devices to protect their plants, but sadly such thefts spoil the enjoyment of ordinary plant lovers, as gardeners can be reluctant to open their gardens to people they do not know.

There are of course happier gardening tales to tell. In 2014 a woman paid £1,602 (around $2,500) at an online auction for the privilege of naming a snowdrop cultivar. Raised by Tom Mitchell, the owner of the nursery Evolution Plants, the snowdrop is a seedling of G. *reginae-olgae* subsp. *vernalis* and has dainty green markings on the outer petals. The successful bidder, Caroline Mabbs of Hertfordshire, has named the snowdrop after her 79-year-old father, Peter Gooding. Gooding, who had to attend the funeral of his own father on his thirteenth birthday, feels that snowdrops have a special significance because at the time of his father's death the family had no money for funeral flowers. Gooding remembers that the only flowers at the scene were the snowdrops already growing in the graveyard. He has been planting a different snowdrop cultivar in his garden each year as a way of commemorating his father and now, thanks to his daughter's generosity, will have his very own snowdrop cultivar and a tale to tell.

New cultivars of snowdrops are being found all the time and the latest ones to cause a stir are those with a hint of other colours in their petals. These include G. *elwesii* 'Anglesey Orange Tip' and 'Senne's Sunrise'.[61] Back in 1891 the *Journal of the Royal Horticultural Society* reported on a snowdrop collector's dream find: 'A. D. Webster tells me that the spring before he left Llandegai he found in the Penshyn woods a pink-flowered snowdrop which he transferred to his garden.'[62] This plant disappeared from sight without being authenticated, but since there is a pink-flowered *Leucojum* species, the possibility of a pink flowered *Galanthus* being found is not out of the question. The finder of such a plant really would have a snowdrop story to tell.

Timeline

16–28 million yrs ago	Period of diversification of the family Amaryllidaceae
4th century BC	Greek philosopher and naturalist Theophrastus used the name *Leucoion* to denote what is assumed to be *Galanthus*
1542	Unidentified snowdrop appears as a coloured engraving in Leonhart Fuchs's *De historia stirpium commentarii insignes*
1554	*Galanthus* is included in Pietro Andrea Mattioli's *Medici senensis commentarii in sex libros pedacii Dioscoridis* under the name *Narcissus* The earliest printed illustration of *Galanthus* appears in the 1554 edition
1575	Snowdrops feature in a botanical illustration by Jacques Le Moyne de Morgues
1722	Early appearance of a snowdrop in English poetry, in Thomas Tickell's 'Kensington Gardens'
1735	Linnaeus names and establishes the genus *Galanthus*
1840	Queen Victoria carries a posy of snowdrops on the day of her wedding to Prince Albert
1875	*Galanthus elwesii* is named by Sir Joseph Hooker

1891	The first snowdrop conference is held by the Royal Horticultural Society in London
1952	Galantamine is extracted from *Galanthus woronowii* for the first time
1956	Frederick Stern's major study *Snowdrops and Snowflakes* is published
1997	First Galanthus Gala organized in Cirencester by Joe Sharman of Monksilver Nursery
1999	*The Genus 'Galanthus'* is published by Aaron Davis
2000	Galantamine is licensed for use in the treatment of Alzheimer's disease in the UK
2001	Publication of *Snowdrops: A Monograph of Cultivated Galanthus* by Matt Bishop, Aaron Davis and John Grimshaw
2012	*Galanthus panjutinii* becomes the twentieth snowdrop species to be recognized

References

Introduction

1 Aaron P. Davis, *The Genus 'Galanthus'* (Portland, OR, 1999), p. 86.
2 Arthur H. Church, *Types of Floral Mechanism: A Selection of Diagrams and Descriptions of Common Flowers, Arranged as an Introduction to the Systematic Study of Angiosperms* (Oxford, 1908), p. 17.
3 Elizabeth Lee, *Ouida: A Memoir* (London, 1914), p. 49.
4 Charles de Sercy, *Poesies choisies de Messieurs Corneille, Benserade, de Scudery . . .* (Paris, 1662), p. 240.
5 Catharine Harbeson Waterman Esling, *Flora's Lexicon: An Interpretation of the Language and Sentiment of Flowers* (Philadelphia, PA, 1840), p. 190.
6 See '21 Facts about Blackbirds', www.rspb.org.uk.
7 Roy Vickery, *The Oxford Dictionary of Plant-lore* (Oxford, 1995), p. 355.
8 Oliver Wyatt, 'The Gardens of Maidwell Hall', *Journal of the Royal Horticultural Society*, LXXXI (1956), pp. 294–302.
9 Roy Lancaster, 'Foxgrove Plants', *The Garden*, CXXXVII/2 (2012), p. 76.
10 Samuel Arnott, 'The Fair Maids of February: The Snowdrop – its History, Literature and Botany', *Transactions and Journal of the Proceedings of the Dumfriesshire and Galloway Natural History and Antiquarian Society* (1904), pp. 339–50.
11 E. Charles Nelson, 'James Atkins and his Plants', *The Plantsman*, X/1 (2011), p. 54.
12 Christopher Lloyd, *Cuttings: A Year in the Garden with Christopher Lloyd* (London, 2008), pp. 33–4.
13 Mirabel Osler, *In the Eye of the Garden* (New York, 1993), p. 168.
14 R. Browning and E. B. Browning, *The Letters of Robert Browning and Elizabeth Barrett Browning, 1845–1846, with Portraits and Facsimiles, in Two Volumes*, vol. I (London, 1900), p. 15.
15 Reginald Farrer, *In a Yorkshire Garden* (London, 1909), p. 5.
16 *The National Plant Collections Directory 2014: Plant Heritage* (London, 2014), p. 70.
17 Barbara Tiffany, *Hunting the Wild Galanthus in the Republic of Georgia* (Hardy Plant Society Mid-Atlantic Group newsletter), XXVI/3 (2012) p. 5.

1 Among Trees and Rocks

1 Richard Mabey, *Flora Britannica* (London, 1996), p. 341.
2 M. D. Lledó, A. P. Davis, M. B. Crespo et al., 'Phylogenetic Analysis of *Leucojum* and *Galanthus* (Amaryllidaceae) Based on Plastid matK and Nuclear Ribosomal Spacer (ITS) DNA Sequences and Morphology', *Plant Systematics and Evolution*, CCXLVI (2004), pp. 223–43.
3 Richard Salisbury and William Hooker, *The Paradisus Londinensis; or, Coloured Figures of Plants Cultivated in the Vicinity of the Metropolis* (London, 1805–8), vol. II, available at www.biodiversitylibrary.org, accessed 12 July 2014.
4 John Gilbert Baker, *Journal of the Linnean Society of London, Botany*, XVI (1878), p. 678.
5 J. McNiell, F. R. Barrie, H. M. Burdet et al., 'Regnum Vegetabile', *International Code of Botanical Nomenclature (Vienna Code)* (2006), p. 146.
6 Matt Bishop, Aaron Davis and John Grimshaw, *Snowdrops: A Monograph of Cultivated 'Galanthus'* (Maidenhead, 2001), p. 2.
7 Guido Aschan and Hardy Pfanz, 'Why Snowdrop (*Galanthus nivalis* L.) Tepals have Green Marks', *Flora: Morphology, Distribution, Functional Ecology of Plants*, 201 (2006), pp. 623–32.
8 Arthur H. Church, *Types of Floral Mechanism: A Selection of Diagrams and Descriptions of Common Flowers, Arranged as an Introduction to the Systematic Study Angiosperms* (Oxford, 1908), pp. 17–31.
9 Nevin Ferda Sahin, 'Pollen Morphology of *Galanthus elwesii* Hooker Amaryllidaceae', *Pakistan Journal of Botany*, XXXII (2000), pp. 5–6.
10 Rosemary J. Newton, Fiona R. Hay and Richard H. Ellis, 'Seed Development and Maturation in Early Spring-flowering *Galanthus nivalis* and *Narcissus pseudonarcissus* Continues Post-shedding with Little Evidence of Maturation in Planta', *Annals of Botany*, CXI (2013), pp. 945–55.
11 Gennadij Budnikov and Vladimir Kricsfalusy, 'Bioecological Study of *Galanthus nivalis* L. in the East Carpathians', *Thaiszia – Journal of Botany*, IV (1994), pp. 49–75.
12 Nuran Ekici and Feruzan Dane, 'Calcium Oxalate Crystals in Vegetative and Floral Organs of *Galanthus* sp. (Amaryllidaceae)', *Asian Journal of Plant Sciences*, VI (2007), pp. 508–12.
13 Leonhart Fuchs, *Das Kräuterbuch von 1543 Von weiß Hornungßblumen.* Cap. CLXXXV Abb 274 (p. 489): Weiß Hornungßblum mit dem samen (*cclxxiii*).
14 Pierre Edmond Boissier, *Flora orientalis* (Geneva, 1882), vol. V, pp. 144–6.
15 John Gilbert Baker, *Handbook of the Amaryllideae* (London, 1888), pp. 16–18.
16 Frederick C. Stern, *Snowdrops and Snowflakes: A Study of the Genera 'Galanthus' and 'Leucojum'* (London, 1956).
17 Aaron P. Davis, *The Genus 'Galanthus'* (Portland, OR, 1999), p. 28.
18 Ibid., p. 122.
19 Ibid., p. 129.

20 Jurij Ivanovich Koss, *Galanthus angustifolius* Botanicheskie Materialy Gerbariya Botanicheskogo Instituti Imeni V. L. Komarova Akademii Nauk S S S R. 14 (1951), p. 134.
21 Graham Stuart Thomas, *Colour in the Winter Garden* (London, 1998), p. 160.
22 Wolfgang Kletzing, 'Travels in Search of Wild Snowdrops', in *Daffodils, Snowdrops and Tulips Yearbook, 2006–2007* (London, 2007), p. 68.
23 Thomas, *Colour in the Winter Garden*, p. 163.
24 Joseph Dalton Hooker, '*Galanthus elwesii*', *Curtis's Botanical Magazine*, CI (1875), pp. 166–74.
25 See '*Galanthus fosteri*', www.encyclopaedia.alpinegardensociety.net, accessed 12 July 2014.
26 Bishop et al., *Snowdrops: A Monograph*, p. 74.
27 Ruby Baker, 'Green-tipped Snowdrops', in *Daffodils, Snowdrops and Tulips Yearbook*, p. 41.
28 *Galanthus ikariae*, in Online Atlas of the British and Irish Flora, at www.brc.ac.uk/plantatlas, accessed 1 June 2015.
29 Wolfam Lobin, Christopher Brickell and Aaron Davis, '*Galanthus koenenianus* (Amaryllidaceae), a Remarkable New Species of Snowdrop from N. E. Turkey', *Kew Bulletin*, XLVIII (1993), pp. 161–3.
30 Andrej Pavlovich Khokhrjakov, 'A New Snowdrop from the Caucasus', *Byul Mosk Obshch Ispytat Prirody Otd Biol* (*Bulletin of the Moscow Society for Nature Investigation*), LXVIII (1963), pp. 140–41.
31 Liubov Manucharovna Kemularia-Nathadze, 'Galanthi Generis Species Novae in *Flora Georgica* Descriptae', *Zametki Sistematike i Geografii Rastenii*, XIII (1947), p. 6.
32 Stern, *Snowdrops and Snowflakes*, p. 42.
33 Dmitriy Zubov and Aaron Davis, '*Galanthus panjutinii* sp. nov.: A New Name for an Invalidly Published Species of *Galanthus* (Amaryllidaceae) from the Northern Colchis Area of Western Transcaucasia', *Phytotaxa*, L (2012), pp. 55–63.
34 Aaron Davis and Christopher Brickell, '*Galanthus peshmenii*: A New Snowdrop from the Eastern Aegean', *New Plantsman*, I (1993), pp. 14–19.
35 Dave Mountfort, 'Newcastle AGS/SRGC Show, 2013', www.alpinegardensociety.net, accessed July 2014.
36 Davis, *Genus Galanthus*, pp. 188–92.
37 Bishop et al., *Snowdrops: A Monograph*, p. 55.
38 Ibid., pp. 105–14.
39 See A. Davis, '*Galanthus plicatus*', IUCN Red List of Threatened Species Version 2014.3, www.iucnredlist.org, accessed 1 June 2015.
40 Bishop et al., *Snowdrops: A Monograph*, p. 163.
41 James Allen, 'Snowdrops', *Journal of the Royal Horticultural Society of London* (1891), p. 179.
42 Davis, *Genus Galanthus*, pp. 95–105.
43 Christopher Brickell, 'Some Caucasian and Turkish Snowdrops in the Wild', in *Daffodils, Snowdrops and Tulips Yearbook*, p. 53.

44 Aleksandr Vasiljevich Fomin and Woronow, *Identification Key to the Plants of the Caucasus and Crimea*, vol. I (1909), p. 280.
45 Aaron Davis and Neriman Ozhatay, '*Galanthus trojanus*: A New Species of *Galanthus* (Amaryllidaceae) from North-western Turkey', *Botanical Journal of the Linnean Society*, CXXXVII (2001), pp. 409–12.
46 Gulem Irem Kaya et al., 'Antiprotozoal Alkaloids from *Galanthus trojanus*', *Phytochemistry*, IV (2011), pp. 301–5.
47 Agnia Losina-Losinskaja, *Galanthus in Flora* SSSR, IV (Leningrad, 1935), pp. 476–80.
48 Brickell, 'Some Caucasian and Turkish Snowdrops in the Wild', p. 48.
49 Davis, *Genus Galanthus*, pp. 193–6.
50 Bishop et al., *Snowdrops: A Monograph*, pp. 58–61.

2 Purity and Piety

1 Margaret Oliphant, *The Autobiography and Letters of Mrs M.O.W. Oliphant* (New York, 1899), p. 28.
2 Richard Mabey, *Flora Britannica* (London, 1996), p. 421.
3 Frederick Holweck, 'Candlemas', in *The Catholic Encyclopedia: An International Work of Reference on the Constitution, Doctrine, Discipline and History of the Catholic Church* (New York, 1908), p. 46.
4 Mabey, *Flora Britannica*, p. 425.
5 Derek Jarman, *Derek Jarman's Garden* (London, 1995), p. 55.
6 Leslie A. Donovan, ed., *Women Saints' Lives in Old English Prose* (Woodbridge, 1999), p. 45.
7 Kevin Danaher, *The Year in Ireland: Irish Calendar Customs* (Dublin, 1972), p. 38.
8 Matt Bishop, Aaron Davis and John Grimshaw, *Snowdrops: A Monograph of Cultivated 'Galanthus'* (Maidenhead, 2001), p. 338.
9 Mabey, *Flora Britannica*, p. 425.
10 György Buday, *The History of the Christmas Card* (London, 1954), p. 304.
11 See 'World's First Printed Valentine's Card' www.bbc.co.uk/ahistoryoftheworld, accessed July 2014.
12 Nancy Rosin, 'Mother of the American Valentine', *American History*, XL (2005), pp. 62–4.
13 Dorothy Gladys Spicer, *Festivals of Western Europe* (Alcester, Warwickshire, 1973), p. 23.
14 See 'Collectie', www.museumvoorreligieuzekunst.nl, accessed 14 July 2014.
15 '*In Memoriam*: Victorian and Early 20th Century In Memoriam Cards', www.the-lothians.blogspot.co.uk, accessed 2 January 2012
16 Howard Clayton, *Cathedral City: A Look at Victorian Lichfield* (Lichfield, 1981), p. 5.
17 Edna V. Jackson, *Snowdrops: Their Life-history, their Beauty, their Homes, their Message* (London, 1902), p. 73.
18 Roy Vickery, *The Oxford Dictionary of Plant-lore* (Oxford, 1995), p. 354.

19 Mabey, *Flora Britannica*, p. 425.
20 Paula Bartley, *Prostitution: Prevention and Reform in England, 1860–1914* (London, 1999), p. 80.
21 Linda E. Connors and Mary Lu MacDonald, *National Identity in Great Britain and British North America, 1815–1851: The Role of Nineteenth-century Periodicals* (Farnham, Surrey, 2011), p. 117.
22 Mark Rowe, 'Snowdrop Withers but Battle to Curb Guns Lives On', *The Independent*, 2 May 1997.
23 Robert Rolfe, 'Plant Awards, 2003–2004', *Alpine Garden Society Bulletin*, LXXII (2004), p. 391.
24 See 'Dave Unveils Snowdrop Walk', www.mkdons.com/news, 3 March 2014.
25 The National Gardens Scheme, *The Yellow Book 2014* (London, 2014).
26 Aida Huseynova, 'Choral Music in West and Central Asia', in *The Cambridge Companion to Choral Music*, ed. André de Quadros (Cambridge, 2012), p. 172.
27 The Romanian Cultural Centre London, 'The Red and White Spring Ball', www.romanianculturalcentre.org.uk, 28 March 2008.

3 Art and Images

1 Mirabel Osler, *In the Eye of the Garden* (New York, 1993), p. 174.
2 Erika Langmuir, *A Closer Look: Still-life* (London, 2010), p. 18.
3 William T. Stearn, *Botanical Masters* (New York, 1990), p. 7.
4 Wilfrid Blunt, *The Art of Botanical Illustration* (New York, 1951), pp. 3–4.
5 Paul Hulton, *The Work of Jacques le Moyne de Morgues: A Huguenot Artist in France, Florida, and England* (London, 1977), p. 127.
6 Georges Duby and Philippe Aries, eds, *A History of Private Life: Revelations of the Medieval World* (Cambridge, MA, 1993), p. 580.
7 The British Library Board, '*Hortus eystettensis*', www.bl.uk/onlinegallery, 26 March 2009.
8 Mark Mitchell, 'The Low Countries' Use of Flowers in Still-life Paintings', www.markmitchellpaintings.com, accessed 14 August 2014.
9 Langmuir, *A Closer Look*, pp. 34–8.
10 Walter A. Liedtke, *Dutch Paintings in the Metropolitan Museum of Art*, vol. I (New York, 2007), p. 314.
11 Antonia Ridge, *The Man Who Painted Roses: Story of Pierre-Joseph Redouté* (London, 1979).
12 Leslie Parris, 'Philip Reinagle, 1749–1833', www.tate.org.uk, June 1997.
13 Alan Thomas, *Great Books and Book Collectors* (Littlehampton, 1975), pp. 142–4.
14 Handasyde Buchanan, *Thornton's Temple of Flora* (London, 1951), p. 15.
15 J. J. Grandville, Alphonse Karr and Taxile Delord, *The Flowers Personified* [1847], at www.publicdomainreview.org, accessed 14 August 2014.
16 Alois Senefelder, *The Invention of Lithography*, at www.gutenberg.org, accessed 23 July 2014.

17 Christopher Wood, *The Pre-Raphaelites* (London, 1981), pp. 9–10.
18 Sophia Andres, *The Pre-Raphaelite Art of the Victorian Novel: Narrative Challenges to Visual Gendered Boundaries* (Chicago, IL, 2005), p. 65.
19 Euphemia Millais, letter to Mrs Gray, 2 March 1854, in Mary Lutyens, *Millais and the Ruskins* (London, 1967), p. 148.
20 Charles Algernon Swinburne, *Posthumous Poems*, ed. Edmund Gosse and Thomas J. Wise (London, 1917), p. 112.
21 Virginia Surtees, *The Paintings and Drawings of Dante Gabriel Rossetti (1828–1882): A Catalogue Raisonné*, vol. I (London, 1971), p. 172.
22 Oscar Lovell Triggs, *The Arts and Crafts Movement* (New York, 2009), p. 185.
23 Julie M. Johnson, *The Memory Factory: The Forgotten Women Artists of Vienna 1900* (West Lafayette, IN, 2012), pp. 55–111.
24 Anne Karhio, 'Seamus Heaney, Paul Durcan and Hugo Simberg's "Wounded Angel"', *Nordic Irish Studies*, XI/I: *The Island and the Arts* (2012), pp. 27–38.
25 'FOKUS: The Angel', Ateneum Art Museum. www.ateneum.fi, accessed 5 October 2014.
26 Richard Shone, *The Art of Bloomsbury: Roger Fry, Vanessa Bell and Duncan Grant* (Princeton, NJ, 1999), pp. 137–8.
27 Christine Cariati, 'Winifred Gill and the Omega Workshops', www.venetianred.net, 16 December 2009.
28 Denise Ortakales, 'Ida Rentoul Outhwaite (1888–1960)', www.ortakales.com, 24 August 2002.
29 Chen Kelun, *Chinese Porcelain: Art, Elegance, and Appreciation* (San Francisco, CA, 2004), p. 3.
30 Sarah Richards, *Eighteenth-century Ceramics: Products for a Civilised Society* (Manchester, 1999), pp. 23–6.
31 'About *Flora Danica*', www.kb.dk (The Royal Library, Denmark), accessed 31 July 2014.
32 'The *Flora Danica* Service', www.royalcollection.org.uk, accessed 31 July 2014.
33 Gordon Campbell, ed., *The Grove Encyclopedia of Decorative Arts* (Oxford, 2006), vol. I, p. 460.
34 Oliver Watson, *British Studio Pottery: Victoria and Albert Museum Collection*, exh. cat., Victoria & Albert Museum, London (1990).
35 Fraser Street, *Moorcroft: The Phoenix Years* (London, 2000).
36 'Little Sparta Goes a Long Way in Poll on Scotland's Greatest Art', www.scotsman.com, 4 December 2004.
37 Hans van Lemmen, *Victorian Tiles* (Oxford, 1999), p. 17.
38 Christopher Morley, *Christopher Dresser's Decorative Design* (London, 2010), p. 256.
39 Hans van Lemmen, ed., *Fired Earth: 1,000 Years of Tiles in Europe* (Ilminster, 1991).
40 'St Paul's Jarrow', www.stpaulschurchjarrow.com, accessed 14 July 2014.
41 Rosemary C. LoDato, *Beyond the Glitter: The Language of Gems in Modernista Writers Rubén Darío, Ramón del Valle-Inclán, and José Asunción Silva* (Lewisburg, PA, 1999), p. 56.
42 'Vase aux perce-neige', www.art-lor-marteau.blogspot.fr, 26 April 2012.

43 'Tenture de la Dame à la licorne', www.musee-moyenage.fr, 16 July 2014.
44 Jimmy Page and Paul Reeves, 'They Shook Me: Pre-Raphaelites Victorian Avant-garde at Tate Britain II', www.tate.org.uk, 29 August 2012.
45 'Millennium Tapestry', www.ufford.onesuffolk.net, accessed 24 July 2014.
46 Roger Berthoud, 'Sutherland, Graham Vivian (1903–1980)', *Oxford Dictionary of National Biography* (Oxford, 2004).

4 Words and Music

1 Rebecca Gethin, 'Language Garden', *THE SHOP: A Magazine of Poetry*, XLIII (2013), p. 70.
2 'Top Names of the 1900s', www.ssa.gov/oact/babynames, accessed 15 August 2014.
3 George Arnett, 'The Top 100 Baby Names in England and Wales in 2013', www.theguardian.com, 15 August 2014.
4 Hilda Ellis Davidson and Anna Chaudhri, eds, *A Companion to the Fairy Tale* (Martlesham, Suffolk, 2006), p. 75.
5 Alex Ben Block and Lucy Autrey Wilson, eds, *George Lucas's Blockbusting: A Decade-by-decade Survey of Timeless Movies Including Untold Secrets of their Financial and Cultural Success* (New York, 2010), p. 206.
6 R. B. Shaberman, 'George MacDonald and Lewis Carroll', *North Wind*, I (1982), pp. 10–30, at www.snc.edu/northwind, accessed 14 August 2014.
7 George Edward Woodberry, *Nathaniel Hawthorne*, in *American Men of Letters* (Boston, MA, 1902), p. 207.
8 Beatrix Potter, *Beatrix Potter's Journal* (London, 2011).
9 Marta McDowell, *Beatrix Potter's Gardening Life* (London, 2013), p. 285.
10 Ibid., p. 162.
11 '100 Interesting Facts', www.merseyferries.co.uk, accessed 14 August 2014.
12 Georgie Butcher, 'Olympic Torch Route, Day 60: Home of a Rodin and an Unlikely Avalanche', www.theguardian.com, 17 July 2012.
13 Lee Child, *Jack Reacher's Rules* (London, 2012), p. 23.
14 Ed Conway, *The Summit: The Biggest Battle of the Second World War – Fought Behind Closed Doors* (London, 2014).
15 Philip Eden, 'Operation Snowdrop: The Cold Winter of 1955', www.weatheronline.co.uk, accessed 17 August 2014.
16 Beverly Seaton, *The Language of Flowers: A History* (Charlottesville, VA, 2012), p. 62.
17 Mary Wortley Montagu, *Letters of the Right Honourable Lady Mary Wortley Montagu* (London, 1763), pp. 121–2.
18 Aubry de La Mottraye, *A. de La Motraye's Travels through Europe, Asia, and into Part of Africa . . .* (London, 1723).
19 Ronan Deazley, 'Commentary on the Engravers' Act (1735)' (2008), in *Primary Sources on Copyright (1450–1900)*, ed. L. Bently and M. Kretschmer, at www.copyrighthistory.org, accessed 1 June 2015.

20 Frederic Shoberl, *The Language of Flowers: With Illustrative Poetry* (Philadelphia, PA, 1839), p. 37.
21 H. Friedmann, *The Symbolic Goldfinch: Its History and Significance in European Devotional Art* (Washington, DC, 1946).
22 Gregory Maertz, ed., *George Eliot's 'Middlemarch'* (Peterborough, 2004).
23 Gregory Maertz, ed., *Cultural Interactions in the Romantic Age: Critical Essays in Comparative Literature* (Albany, NY, 1998).
24 Beverly Lyon Clark, *Louisa May Alcott: The Contemporary Reviews* (Cambridge, 2004), p. 98.
25 William McCarthy, 'Anna Letitia Barbauld (1743–1825)', *Oxford Dictionary of National Biography* (Oxford, 2004), online edn, January 2008, www.oxforddnb.com.
26 A. L. LeBreton, *Memoir of Mrs Barbauld, Including Letters and Notices of her Family and Friends* (Cambridge, 1874).
27 William Wordsworth to Alexander Dyce, 10 May 1830, in William Knight, ed., *Letters of the Wordsworth Family*, vol. II (New York, 1907) pp. 428–9.
28 Howard Jacobson, in Vivienne Westwood, 'My God, the Beauty of Gainsborough!', *The Telegraph: Review*, 30 August 2014.
29 Kristen Chancey, 'Mary Robinson' http://users.clas.ufl.edu/pcraddoc/chancey.htm, accessed 5 August 2014.
30 James Bieri, *Percy Bysshe Shelley* (Baltimore, MD, 2008), pp. 519–21.
31 Paul Mariani, *Gerard Manley Hopkins: A Life* (New York, 2008), p. 97.
32 Sarah Parker, 'Olive Custance (1874–1944)', in *The Yellow Nineties*, ed. Dennis Denisoff and Lorraine Janzen Kooistra, www.1890s.ca, 2010.
33 A. Plaitakis and R. C Duvoisin, 'Homer's Moly Identified as *Galanthus nivalis* L.: Physiologic Antidote to Stramonium Poisoning', *Clinical Neuropharmacology*, VI (1983), pp. 1–5.
34 A. Alzheimer, R. A. Stelzmann, H. N. Schnitzlein and F. R. Murtagh, trans., 'An English Translation of Alzheimer's 1907 Paper, about a Peculiar Disease of the Cerebral Cortex', *Clinical Anatomy*, VIII (1995), pp. 429–31.
35 P. I. Russo, A. Frustaci, A. Del Bufalo et al., 'From Traditional European Medicine to Discovery of New Drug Candidates for the Treatment of Dementia and Alzheimer's Disease: Acetylcholinesterase Inhibitors', *Current Medicinal Chemistry*, XX (2013), pp. 976–83.
36 Michael Heinrich and Hooi Lee Teoh, 'Galanthamine from Snowdrop: The Development of a Modern Drug Against Alzheimer's Disease from Local Caucasian Knowledge', *Journal of Ethnopharmacology*, XCII (2004), pp. 147–62.
37 'NICE Technology Appraisals [TA217]: Donepezil, Galantamine, Rivastigmine and Memantine for the Treatment of Alzheimer's Disease, www.nice.org.uk, March 2011.
38 Final Report Summary: SUPROGAL (Sustainable Production of Galanthamine by Both In Vitro and Agricultural Crops of Highly Galanthamine-containing Plants), www.cordis.europa.eu, 17 June 2013.

39 Douglas Lew, *Great Composers in Watercolor* (Bloomington, IN, 2010), p. 94.
40 Elisabeth M. Orsten, *From Anschluss to Albion: Memoirs of a Refugee Girl, 1938–1940* (Cambridge, 1998), p. 20.
41 Michael Musgrave, *A Brahms Reader* (New Haven, CT, 2001), p. 203.
42 Bell Gale Chevigny, ed., *The Woman and the Myth: Margaret Fuller's Life and Writings* (Lebanon, NH, 1976), p. 363.
43 *Musical Herald* (1 September 1897), p. 297.
44 Josceline Dimbleby, *A Profound Secret: May Gaskell, her Daughter Amy, and Edward Burne-Jones* (London, 2012), p. 62.
45 Marco Rocha, 'Electro Band Handful of Snowdrops Returns', www.subterock.com, 16 December 2013.

5 Collectors and Conservation

1 C. S. Lewis, *C. S. Lewis' Letters to Children*, ed. Lyle W. Dorsett and Marjorie Lamp Mead (New York, 1996), p. 39.
2 Naomi Slade, *The Plant Lover's Guide to Snowdrops* (Portland, OR, 2014), p. 10.
3 Joyce Tyldesley, *Hatchepsut: The Female Pharaoh* (London, 1996), p. 145.
4 Thomas Christopher, *In Search of Lost Roses* (London, 1998), p. 81.
5 Douglas Chambers, '"Storys of Plants": The Assembling of Mary Capel Somerset's Botanical Collection at Badminton', *Journal of the History of Collections*, IX (1997), pp. 49–60.
6 Molly McClain, *Beaufort: The Duke and his Duchess, 1657–1715* (New Haven, CT, 2001), p. 213.
7 *Gardeners' Chronicle* (2 April 1842), p. 222.
8 Graham Stuart Thomas, *A Garden of Roses: Watercolours by Alfred Parsons* (London, 1987), p. 68.
9 Audrey Le Lièvre, *Miss Willmott of Warley Place: Her Life and her Gardens* (London, 1980), p. 193.
10 See www.nationalarchives.gov.uk/battles/crimea, 18 September 2014.
11 Edward Augustus Bowles, 'Garden Varieties of *Galanthus*', in Frederick C. Stern, *Snowdrops and Snowflakes: A Study of the Genera 'Galanthus' and 'Leucojum'* (London, 1956), p. 120.
12 Aaron P. Davis, *The Genus 'Galanthus'* (Portland, OR, 1999), p. 193.
13 W. W., 'In Memoriam Frederick William Burbidge', *Journal of the Kew Guild* (1905), p. 327.
14 Frederick William Burbidge, 'An Irish Garden', *Gardeners' Chronicle* (17 June 1893).
15 Frederick Burbidge, *The Gardens of the Sun; or, a Naturalist's Journal on the Mountains and in the Forests and Swamps of Borneo and the Sulu Archipelago* (London, 1880).
16 James Allen, 'Snowdrops', *Journal of the Royal Horticultural Society*, XXIII/2 (1891), p. 180.
17 Frederick William Burbidge, 'Snowdrops', *Journal of the Royal Horticultural Society* vol. XXIII/2 (1891), pp. 191–209.

18 Frederick William Burbidge, 'Autumn-blooming Snowdrops', *The Garden*, XXXIX (1891), p. 243.
19 Frank Harris, *Oscar Wilde* (1916, corrected edn 1938), p. 29.
20 Edward Augustus Bowles, *My Garden in Spring* (London, 1914), p. 48.
21 Jan Marsh, *William Morris and Red House* (London, 2005), p. 60.
22 Matt Bishop, Aaron Davis and John Grimshaw, *Snowdrops: A Monograph of Cultivated Galanthus* (Maidenhead, 2001), pp. 285–302.
23 Frederick C. Stern, *Snowdrops and Snowflakes: A Study of the Genera Galanthus and Leucojum* (London, 1956).
24 Richard Hobbs, 'H. O. Greatorex's Garden: Snowdrop Acre', *Norfolk Gardens Trust Journal* (2012).
25 Brent Elliott, *Victoria Medal of Honour, 1897–1997, The Royal Horticultural Society* (London, 1997).
26 Bishop et al., *Snowdrops: A Monograph*, p. 55.
27 W. Greuter, J. McNeill, F. R. Barrie et al., eds, *International Code of Botanical Nomenclature (Saint Louis Code)* (Königstein, 2000).
28 'Planting in Perak', *Daily Advertiser*, Singapore (22 May 1891).
29 Gunnar Seidenfaden, Jeffrey J. Wood and Richard Eric Holttum, *The Orchids of Peninsular Malaysia and Singapore* (Fredensborg, 1992), p. 708.
30 John Alexander Wilson Gunn, ed., *Benjamin Disraeli: Letters, 1857–1859* (Toronto, 2004), p. 166.
31 Courtney Page, 'New English Roses of 1921', in *The American Rose Annual*, ed. J. Horace McFarland (Harrisburg, PA, 1922), p. 144.
32 Edward Bertram Anderson, *Hardy Bulbs*, vol. I (Harmondsworth, 1964), p. 104.
33 Beth Chatto, *The Damp Garden* (London, 1982), p. 62.
34 Gwynneth Reynolds and Diana Grace, *Benton End Remembered* (London, 2002), p. 150.
35 Beverley Nichols, *Merry Hall* (Portland, OR, 1998), p. 170.
36 Beverley Nichols, *Garden Open Today* (Portland, OR, 2002), pp. 44–5.
37 Beverley Nichols, *Down the Garden Path* (Woodbridge, 1983), pp. 78–9.
38 Ibid., p. 79.
39 Julia Brittain, *Plant Lover's Companion: Plants, People and Places* (Newton Abbot, Devon, 2006), p. 81.
40 Daphne Chappell, 'Born of a Garden', *Cottage Gardener*, XLVII (1994) pp. 9–11.
41 Bishop et al., *Snowdrops: A Monograph*, p. 336.
42 Tony Copsey, 'John E. Morley', www.suffolkpainters.co.uk, accessed 15 October 2014.
43 John Morley, *North Green Snowdrops 2012 Catalogue* (Beccles, 2012).
44 Mary Keen, 'That Old White Magic', *The Telegraph*, 11 February 2006.
45 Robert Rolfe, *The Alpine Gardener supp Awards of the Joint Rock Garden Plant Committee* (Pershore, 2010–11), p. 5.
46 Audrey Woods, 'Snowdrops Give Way to Garden Paradise', *Ellensburg Daily Record*, 13 April 2001.
47 Bishop et al., *Snowdrops: A Monograph*, p. 84.

48 S. Berkov, J. Bastida, B. Sidjimova et al., 'Alkaloid Diversity in *Galanthus elwesii* and *Galanthus nivalis*', *Chemistry and Biodiversity*, VIII (2011), pp. 115–30.

49 E. Y. Nakasu, S. M. Williamson, M. G. Edwards et al., 'Novel Biopesticide Based on a Spider Venom Peptide Shows no Adverse Effects on Honeybees', *Proceedings of the Royal Society B: Biological Sciences*, CCLXXXI (2014), p. 1787.

50 *Gardeners' Chronicle*, II (12 March 1842), p. 172.

51 Anthony Florian Madinger Willich and Thomas Cooper, *The Domestic Encyclopedia; or, A Dictionary of Facts and Useful Knowledge, Chiefly Applicable to Rural & Domestic Economy: with an Appendix* . . ., Katherine Golden Bitting Collection on Gastronomy (Library of Congress) (Philadelphia, PA, 1821), p. 259.

52 Erasmus Darwin, *The Botanic Garden Part II* (London, 1791).

53 Jan Willem Moll and Johannes Cornelis Schoute, *Phytography as a Fine Art* (Leiden, 1934), p. 507.

54 Davis, *The Genus 'Galanthus'*, pp. 52–3.

55 See www.cites.org, accessed 15 August 2014.

56 Tom Mitchell, *Snowdrops and Hellebores 2012* (Catalogue of Evolution Plants), p. 9.

57 A. Entwistle, S. Atay, A. Byfield et al., 'Alternatives for the Bulb Trade from Turkey: A Case Study of Indigenous Bulb Propagation', *Oryx*, XXXVI (2002), pp. 333–41.

58 *The National Plant Collections Directory 2014: Plant Heritage* (London, 2014), p. 70.

59 'Stealing from Mother Nature', www.news.bbc.co.uk, 22 February 2000.

60 See Susanne Mayer, 'Ein weißes Feld', in www.zeit.de, 2011.

61 Valentin Wijnen, *International Rock Gardener*, XXXVII (2013), p. 7.

62 James Allen, 'Snowdrops', *Journal of the Royal Horticultural Society*, XXIII/2 (1891), p. 180.

Further Reading

Bishop, Matt, Aaron Davis and John Grimshaw, *Snowdrops: A Monograph of Cultivated Galanthus* (Maidenhead, 2001)
Cox, Freya, *A Gardener's Guide to Snowdrops* (Marlborough, 2013)
Davis, Aaron P., *The Genus 'Galanthus'* (Portland, OR, 1999)
Dijk, Hanneke van, *Galanthomania* (Arnhem, 2011)
Jackson, Edna V., *Snowdrops: Their Life-history, their Beauty, their Homes, their Message* (London, 1902)
Mabey, Richard, *Flora Britannica* (London, 1996)
Slade, Naomi, *The Plant Lover's Guide to Snowdrops* (Portland, OR, 2014)
Stern, Frederick C., *Snowdrops and Snowflakes: A Study of the Genera Galanthus and Leucojum* (London, 1956)
Waldorf, Gunter, *Snowdrops* (London, 2011)

Associations and Websites

ALGEMEENE KONINKLIJKE VEREENIGING VOOR BLOEMBOLLENCULTUUR
(KAVB, Royal General Bulb Growers' Association, Netherlands)
www.kavb.nl
The international registration authority for *Galanthus* cultivars

ALPINE GARDEN SOCIETY
www.alpinegardensociety.net

COTTAGE GARDEN SOCIETY: SNOWDROP GROUP
www.thecottagegardensociety.org.uk

HARDY PLANT SOCIETY: GALANTHUS GROUP
www.hardy-plant.org.uk/galanthus

INTERNATIONAL BULB SOCIETY
www.bulbsociety.org

NATIONAL GARDENS SCHEME
www.ngs.org.uk

NORTH AMERICAN ROCK GARDEN SOCIETY
www.nargs.org

PACIFIC BULB SOCIETY
www.pacificbulbsociety.org

PLANT HERITAGE: NATIONAL COUNCIL FOR THE CONSERVATION OF PLANTS AND GARDENS
www.plantheritage.com

SCOTTISH ROCK GARDEN CLUB
www.srgc.org.uk

Acknowledgements

Gardeners on the whole are generous people and I have been fortunate during the writing of this book to encounter many snowdrop admirers who have been generous not just with their knowledge and plants but also with friendship and cake. Members of the Scottish Rock Garden Club Forum have been invaluable in all these regards. Special thanks go to National Collection Holder Steve Owen and to the staff of the Lindley Library for their assistance. Thank you to Richard Fennell for chauffeuring me to snowdrop gardens and spotting many snowdrops in art galleries. Ashley and Jonathan Harland have assisted with computer problems and have been emotionally supportive as always.

Photo Acknowledgements

The author and publishers wish to express their thanks to the following sources of illustrative material and/or permission to reproduce it:

Alamy: pp. 10 (The Natural History Museum), 190 (Andrea Jones Images); © Olga Bondareva: pp. 39, 40, 44, 49, 52, 54, 56; © The British Library Board: p. 64; Sir Peter Erskine, Cambo House: p. 26; © The Fitzwilliam Museum, Cambridge: p. 91; Garden World Images: p. 16 (Steffen Hauser); Gail Harland: pp. 6, 8, 18, 22–3, 31, 33, 34 (top), 43, 62, 75, 79, 83, 121, 126, 133, 145, 160, 176; KHM-Museumsverband: p. 95; Mary Evans Picture Library: p. 134; Howard Rice Garden Photography: pp. 20, 45, 46, 158, 179; Science Photo Library: p. 34 bottom (SCIMAT); Victoria & Albert Museum, London: pp. 88, 98, 105.

Index